ENTERPRISE MOBILITY
STRATEGY & SOLUTIONS

ENTERPRISE MOBILITY STRATEGY & SOLUTIONS

Rakesh Patel

PARTRIDGE
A Penguin Random House Company

To order additional copies of this book, contact
Partridge India
000 800 10062 62
orders.india@partridgepublishing.com

www.partridgepublishing.com/india

CONTENTS

Section 2 – Enterprise Mobility in the Workplace

Chapter 3:   Enterprise Mobility Team
and a Mobile Workforce

Section 3 – The Scope of Enterprise Mobility

Section 4 – Other Aspects of Enterprise Mobility

PREFACE

With smart phones turning the preferred devices and mobile becoming one of the most preferred medium for computing, enterprises find prudence and practicality in adopting and accommodating mobile technologies. And why would they not, after all this could also lead them to improved operational efficiencies and better revenues. Which business would not want it anyway?

Having said so, enterprise mobility is not obviously something that businesses can just decide and have everything set up accordingly in no time. It requires a great effort, a good strategy, substantial investments and lots more to avail and bring everything down to the ease of mobility for customers as well as employees. Clearly there will be complexities and elusiveness for enterprises since Enterprise Mobility Management (EMM) is still an emerging discipline; and they are very few best approaches and examples to follow.

Now as few are the best practices or benchmarks to follow, there's even a lesser availability of comprehensive guidebooks on enterprise mobility, which provide end-to-end clarity on the subject. And since someone has to make the move, I took it on myself to make an attempt, to document everything (well, almost everything) about enterprise mobility that could help IT decision makers, and enterprisers (like me) to sail through the tide.

The book walks you through the various aspects of enterprise mobility. It has four sections which are further divided into 13 chapters. The first section is comprised of the Introduction and the popular trends related with Enterprise Mobility. It even talks about the organizations that have joined the mobility revolution and the challenges they have faced. The second section is about the enterprise mobility in the workplace. It covers the pros and the cons of having a mobile workforce, along with discussing the mobile devices and platforms; and devising the enterprise mobility strategy. The third section talks about the scope of enterprise mobility covering topics like enterprise mobility management and enterprise mobility solutions. The fourth and the final section, speaks about the important aspects like Mobile Device Management (MDM), BYOD, Mobile Security, and Mobile Business Intelligence. It also features a chapter about the future of Enterprise Mobility.

It took months of hard work and effort to finish and furnish this guide cum Handbook for enterprise mobility. However, this book would not have been possible without the support of my family and my team at work. I am sincerely grateful to them for their patience and support. Writing this book has also helped me gain a better insight of the subject, I anticipate that it would equally benefit the readers, particularly the CIOs, CTOs and IT decision makers.

Author Bio

Rakesh Patel is the CEO and Founder of Space-O Technologies. In his long career of 2 decades, Rakesh has played key roles in diverse areas of the software business assuming responsibilities as System Analyst and Design Architect, Project Manager, Functional Consultant, Head of Business Development and a Business Unit Leader.

Rakesh has been into Mobile App space from last 6 years now and has defined successful business models for the clients that worked flawlessly for them. Rakesh has set up 2 state-of-art Development Centers each in India and Russia; and contributes to the overall vision of the organization as a mentor.

Apart from his role as an entrepreneur, Rakesh is also an author, speaker, and role model to many aspiring entrepreneurs. His unique ideas and vast experience have helped his organization, clients, stakeholders and many others to thrive and achieve a progressive level.

SECTION 1 –

The Mobility Revolution

CHAPTER 1

Introduction to Enterprise Mobility and Management

What is Enterprise Mobility?

Let's start with the basics. One hears the terms enterprise mobility, m-enterprise and mobility all the time, but what does the term enterprise mobility exactly mean? According to the McKinsey Quarterly, Enterprise mobility is the deployment of mobile devices of all kinds such as smart phones and tablets as well as related IT staff and infrastructure, network and applications by a company for any of the following three purposes:

- To enhance their engagement with their end customer
- To drive productivity in their internal business processes
- To improve the productivity of their employees

Enterprise Mobility essentially allows a business' employees to have instant access to vital personal and work-related data via mobile applications, Access is available around the clock regardless of where the employee might be at the time. There has been a marked change in the way people work since the introduction of the first iPhone in 2007. If a business wants to remain in the running, investing in enterprise mobility to make applications accessible via mobile devices is an inevitable, yet potentially transformative expense.

What is the Driving Force Behind the Surge in Enterprise Mobility?

An interesting question that might be raised here is what is driving enterprise mobility? Is it simply the development and availability of new technologies such as smart phones or laptops or are there other factors? While the ubiquity of smart phones and the rapid adoption of tablets are important factors, there are a number of other factors spurring enterprise mobility. The primary reasons are as follows:

The deployment of the 4G network across North America is a critical enabler of enterprise mobility because it truly enables applications such as videos and remote access into corporate networks over mobile-based Virtual Private Networks to become a 'reality'.

- The consumerization of IT. Work models that employees are now using in their personal lives, they are also bringing with them to the work are huge drivers as well.
- There has been considerable progress in security and endpoint security, two aspects that previously made enterprises suspicious of mobility. CIOs today are much more comfortable with the secure mobile devices have gotten.

Finally, there has been a macro trend towards jobless recovery, which is the economic phenomenon where a macro-economy experiences growth from maintaining or decreasing employment. Executives face tremendous pressure to increase employee productivity because they have to be able to do more with less human capital in terms of the number of employees and hence mobility comes up as a feasible solution.

According to latest research, there is speculation that every year, there will be 30% growth in the utilization of mobile applications in business enterprises. Thanks to devices such as the iPhone, BlackBerry and Windows Phone in addition to tablets, the business world is witnessing an unparalleled increase in employee productivity and communication. This has been possible because having critical business information readily available is a potent resource especially but not restricted to the areas of sales and marketing.

Mobile access to company software and information leads to improved mobile productivity. This has the potential to translate into a greater return on investment for businesses that invest in enterprise mobility in addition to more sales and enhanced communication flow.

What is the Significance of Enterprise Mobility?

Timely access to information is critical to the efficient functioning of a business. Enterprise mobility increases productivity by enabling access to information precisely when it is needed by providing employees with mobile, immediate communication and access opportunities. Enterprise mobility can also provide critical customer related information at the point of sale, thereby increasing the ability of employees in sales to respond to their client's needs quickly and efficiently. Devoting resources to enterprise mobility is, therefore, not just a cost but also an investment in the company's future and is the next logical step in any Business evolution.

Secondly, enterprise mobility can circumvent the expense of purchasing laptops for all Off-site marketing and sales teams. Employees can update consumer information on the go and have access to all required information without costly equipment. Mobile devices such as smart phones have all the features of a sophisticated Customer Relation Management application with the added benefits of being easier on the pocket (especially because of the Bring Your Own Device BYOD trend) and easier to carry while travelling to distant field sites.

Thirdly, having access to critical information anytime, anywhere with a mere finger tap can increase business efficiency and yield high Make plural on Make plural and follow with a semi-colon instead of a comma, allowing enterprises to surpass their competition -- specially when it comes to the aggressive areas of sales and marketing.

Finally, enterprise mobility ensures smoother workflow by allowing more to be done in less time. Mobile devices can be synchronized with cloud-hosted applications resulting in uninterrupted access to communication and data on the go. Whether it is reading and responding to emails or giving the green light to an employee's proposal -- then things can be done remotely. As a result, efficiency is sure to increase.

What are the challenges of Enterprise Mobility?

Mobilizing an enterprise may be a great idea but that certainly does not mean that it is free of challenges. Broadly speaking, the challenges facing any enterprise wishing to become more mobile make plural into one of four categories.

1. **Living up to the Hype:** You will routinely come across phrases like Italicize or put quotes around this phrase Italicize or put quotes around this phrase. The text will inevitably be followed by impressive stats and figures on how a business experienced a complete turnaround by investing in technology XYZ. However, statistics can sometime paint a rosier picture should read: than is the actual truth. This does not mean that application developers are deliberately communicating false information to hoodwink enterprises into purchasing their mobility solutions, but the fact of the matter is

that these statistics, when quoted out of context, can be somewhat misleading. While there is no denying that there has been a huge increase in the number of organizations deploying smart phones and tablets in business activities, does this technology dissemination Place (especially in North America and Europe) in parenthesis. In North America and Europe always imply a causal impact on a business's efficiency? At this juncture, it is critical to stop and ponder over what mobility Omit 'in fact' means for business in real value terms.

2. **How do you make a case for 'investing in' enterprise mobility?** Justification is conducting a Hyphenate analysis to convince stakeholders of the business of the of an investment. The current recession has forced most enterprises to cut expenses to the bare bones. In this climate of austerity, stakeholders will need hard facts to be willing to spend on enterprise mobility. Chapter 3 offers suggestions to...help you, as a Chief Information Officer, make a solid case for the introduction of expansion of... in your End sentence here. Come prepared to answer questions pertaining to this does not need to be capitalized. As well as the functionality of your proposal.

3. **Keeping abreast with new technology:** With R&D departments working around the clock, it's to no surprise that new technology is being introduced in the market almost daily. The sheer expanse of mobile technology, with as well as its accompanying tech-lingo and jargon, can have quite a dizzying effect on the should read not-so-tech-savvy. Take telling MEAP apart from MDAP and MDM, for instance. Moreover, some of these abbreviations can overlap with others adding to the confusion. The matter boils down to finding

the right solution to meet your business' goals without unnecessarily getting lost in the maze of acronyms and technology.

4. **Keeping up with rapid change:** Since new mobile devices are entering the market every day, it will not take long for today's innovative technology to become obsolete Remember when having the Nokia 3310 made you the envy of everyone? Fast forward a few years and it looks ancient compared to today's Android and Windows smart phones.

For businesses it means additional costs of continually updating devices and software. Moreover, not only are hardware and software changing, but....and software are changing, but a lot is happening to change business processes themselves; making for a ....for a complex milieu.

Enterprise Mobility as an Ecosystem

At its core, a mobility ecosystem is the interaction between people (workers and customers), processes/operations, technology consisting of assets such as networks and devices, as well as systems such as databases and applications.

Recall your lesson on ecosystems from secondary school science. An ecosystem is defined as the complex of a community of organisms and its environment functioning as an ecological unit. So an ecosystem essentially comprises the Sun as the main energy source, with plants, herbivores, carnivores and microbes all depending on one another for sustenance. There are energy flows and efficiency losses involved at each stage. Thinking of a mobility ecosystem in the same manner can prove to be a helpful exercise.

The objective of any enterprise mobility management firm should be to create a customer-centered mobility experience by flawlessly integrating the best mobility technologies and services via cloud computing. This involves working in tandem with technology providers, global system integrators, mobile network operators and mobile device manufacturers.

Working in unison, all strategic partners provide value and return on investment to each participant of the enterprise mobility value chain. The customers also benefit from superior service and convenience.

What is enterprise mobility management?

Hopefully, the previous sections have established the value and importance of enterprise mobility. It should also be...obvious that a mobility ecosystem has several partner and is strife with potential challenges. Therefore, there is a crucial need for some agent(s) to steer the entire system to optimize its gains. This is where the emerging discipline of ENTERPRISE MOBILITY MANAGEMENT (EMM) comes into the picture. EEM encompasses all the personnel, processes and technology that work together to manage the increasing range of mobile devices such as smart phones and tablets, wireless networks and associated services to enable the broad use of mobile computing in today's business environment.

According to a survey, 67% of CIOs and Information Technology Professionals are of the opinion that enterprise mobility will have as dramatic an impact on their businesses as the Internet had in the 1990s. This has led to a sharp rise in the demand for professionals that can manage enterprise mobility especially since more and more workers are bringing their personal mobile devices such as smart phones and tablets to work and are seeking support for optimizing their efficiency in the workplace. A survey by the cube lab[1] of various enterprises revealed that 54% of workers use company-issued and personal devices, 23% used personal devices only while the remaining 23% used only company issued devices. 84% of the enterprises surveyed provided support for corporate issued and owned equipment, 11% provided support to employee-owned equipment and contracts while the last 5% provided support based on employee preference.

Mobile Computing in a Business Context

There is little doubt that mobility is changing the workforce of the 21st century. Businesses have to think on their feet and weave mobile devices into their IT infrastructure. Recent business practices such as BYOD (Bring Your Own Device), BYOA (Bring Your Own Apps) and BYOP (Bring your own parenthesis), combined with the proliferation of 4G connection across the United States, Asia Pacific and Europe, have had a massive impact on how employees are working. The 9 to 5 office is on its way out as more Gen-Y people step into the job market. The Forrester report[2] states that as of 2013, 37 percentage of employees describe themselves as working from multiple: In fact,

---

[1]   http://www.xcubelabs.com/enterprise-mobility-infographic.php

[2]   http://www.xcubelabs.com/enterprise-mobility-infographic.php

82 percent of workers are making use of mobile apps to work remotely a sizeable majority, 82 percent of these workers to be exact, are making use of mobile apps to work remotely.

The BYOD has its merits in that it frees the employer from providing mobile devices. However, the use of personal devices brings with it a threat to company security. Lost or stolen devices can cause critical business data to end up in the wrong hands; putting the whole business in jeopardy. To address this concern, businesses are deploying MDM or Mobile Device Management as a remedy to manage corporate data saved on employee or business owned devices.

In the event of a personal or Hyphenate these words device going missing, enterprises with MDM software in place can minimize or eliminate the danger of crucial business information ending up in the wrong hands by deleting all the information on the stolen/lost device from a remote location on a central computer. Understandably, some employees. Are hesitant in extending their full support to MDMs since there is no clear policy in place regarding the extent to which the enterprise can access and monitor: personal data on the same device. Mobile Device Management software will need to provide employees with clear evidence of what is and is not accessible in order to gain their acceptance and trust on the matter.

The Bring Your Own App BYOA movement is not free from risks either. Employees might use insecure or substandard apps that might result in the…loss of corporate data. CIOs need to invest in assessing the security of various Hyphenate these words apps and devise a list of apps they find suitable for use with corporate data. In many instances, there is a clear shift away from Hyphenate these words towards apps engineered by the IT decision makers in the enterprise in a way that is tailored to cater to the needs of their employees and the business itself.

Mobile Application Management (MAM) software has also spawned in response to the BYOA and BYOD trend. These types of... software and services provide and control access to bespoke as well as off-the-shelf mobile apps used by the enterprise on company-provided or employee-owned devices. Its distinguishing feature from MDM software is the greater control over mobile apps it provides along with comparatively reduced control of the device. It is preferred by employees over MDM since it discriminates between personal apps and corporate apps. This means that even if corporate data on the employee's device is wiped, the employee's personal data remains preserved. MAM also allows enterprises to regulate what apps the employees' device can carry. The IT department can remotely configure the network's settings in a manner such that apps that it assesses to potentially compromise the security of corporate data can be blacklisted from being installed.

Since the number of individuals switching to smart phones is increasing each day, it is safe to assume that BYOD will become increasingly entrenched as a business practice. It is therefore becoming increasingly important that IT decision makers put their heads together to formulate a policy detailing the extent of the enterprise's power, with reference to access to personal data on employees' phones and tablets, as well as the enterprise's expectations with regards to what applications employees can download without putting the firm at risk.

Once the potential data security issues resulting from BYOD and BYOA are tackled using MDM and MAM software, the business can reap enormous benefits from mobile devices and tablets. These benefits include, but are not limited to, improved efficiency and productivity from mobile app use. Businesses can save resources and time by transferring the processing of day-to-day activities such as leave requests and approvals, credit memoranda and timesheets from paper to mobile apps. There is no denying that mobile computing in the business context has countless benefits as long as proper MDM and MAM software is installed by the IT decision makers in the enterprise.

The Emerging Discipline of Enterprise Mobility Management

The preceding sections provided an overview of the definition of enterprise mobility management as well as the threats and benefits of mobile computing in the business context. Now that the reader has an idea of what the needs of an increasingly mobile enterprise are, we can proceed to describing the domains that come under the emerging discipline of enterprise mobility management.

Development of software makes mobile devices, data and applications more secure and manageable. This empowers the enterprise's employees with the productivity resulting from mobility in a secure and streamlined manner.

The wide access to corporate data becoming possible as a result of enterprise mobility brings with it the...threat of loss, theft and unauthorized access. A business's trade secrets, information on partners, client info and other intellectual property are its core assets. A good enterprise mobility management package must be designed to safeguard confidential corporate data in a manner that is effective, yet easy to use and manage by the company's IT personnel.

As stated in the previous section, BYOD or Bring Your Own Device is a trend that is here to stay. BYOD benefits the employee since it enables him/her to exercise greater freedom in the choice of device, but this democratization of mobility puts IT into a tricky situation. By enabling the enterprise to push, install and uninstall apps on the employee's device as well as the authority to wipe/delete data in the event of a threat, EMM increases confidentiality, control, data protection and security.

The best EMM solutions provide comprehensive, enterprise-wide management. What this means is that data stored in the mobile device must be in a format that complies with the data's eventual endpoint such as laptop PCs. EMM solutions allow that by providing a uniform, unified infrastructure across all devices in an enterprise.

The Business Need for Enterprise Mobility Management

In recent years, due to the advancement of technology, worker productivity has greatly increased. This is clearly because modern equipment enables employees to work on the go and thus deliver better results. Similarly, it is also obvious that workers would perform better if they were able to use business or company data.

It seems that the introduction of the first generation iPhone in 2007 was the pivotal event that opened up possibilities for workers to work anywhere at any time and that too, without the help of a desktop computer. However, smart phones cannot be said to be entirely accountable for the rapid increase in productivity that certain companies enjoy, or be held responsible for the ability of employees to work on the go. In fact, in order to comprehend how workers are able to perform on the go, it is important to evaluate a number of aspects such as the information technology atmosphere, mobile operating systems, the training such workers receive and the applications they choose to use.

The use of applications to (apps) is important, as these are software programs that are tailored with specific features that allow workers to perform with better efficiency, both in the workplace and outside it. Because of the invaluable help that different applications provide to corporations, many businesses have work functions that revolve around the use of apps. Furthermore, businesses of a large variety have come to acknowledge the usefulness of mobile applications.

The evidence for the advantages of technology such as smart phones is given by the review of a new study by EMC2. According to the study government workers are able to work worth an extra nine hours due to the use of mobile equipment which results in a possible savings of $28 billion per year.

Similarly, a poll done by the Fierce Government IT showed that employees from all sector exported they worked an additional 3 to 7 hours per week with the help of enterprise mobility This poll also revealed another piece of information; 92 percent of workers using both a smart phone and a tablet felt their performance level was increased but only 78 percent of workers using only a smart phone considered their performance to be better.

Nevertheless, there are certain disadvantages to these devices that threaten to override its benefits. They include the following:

- The extent of safety applications provided
- The differences between personal computer and mobile application versions
- Complicated methods of usage
- Simultaneous editing and process control over crucial corporate data

In order for employees to avoid problems caused by these factors, both the modern technology and corporate sectors must work together and come up with a solution that enables the usage of applications in a safer and more user friendly environment.

The advent of cell phones followed by the introduction of smart phones in 2007, not only revolutionized how the world communicated but also how the world conducted business. A description of the expanse of mobile phone usage puts things in perspective. As recent research has uncovered, 234 million Americans aged 13 and above use a mobile device and 65 million own a smart phone. Of these people, approximately one third of all subscribers can be classified as business users.

2007 brought with it the iPhone the world's first smart …phone, which was unlike any communication device known to humankind. A smart phone differed from your average cell phone because of several core features. It boasted PC functionalities complemented by features such as email, a mini internet browser, external USB options, wide screen, Global Positioning System capability and a huge memory capacity compared to the standard wares available in the market. Moreover, that was not the end of the smart phone's high-end features. Users could also install various applications, tools and programs to personalize their smart phones to their needs and make them extremely versatile devices.

It is no wonder…that these smart phones took the world by storm. The demand was so strong that various manufacturers introduced several more devices with a variety of operating systems out of which the OS of Apple's iPhone and Google Android have the lion's share of the market. Research in Motion's Blackberry was originally perceived as more tailored to the businessperson's needs, but over time, its share in the market has dwindled.

Consumers began to realize the benefits of using these savvy devices to improve processes such as checking their email accounts on the go. Employers too, sought to tap the enormous potential of using mobile devices to enhance their enterprise's efficiency, employee productivity, improve customer relations, increase sales, streamline decision-making, enhance employee satisfaction by enabling greater flexibility, gain competitive edge and reduce the costs of doing business. However, they were also quick to grasp that supporting such a wide array of device types and operating systems would be far from simple and would involve security risks and high setup and maintenance expenses. This necessitated a turnkey method of device management. 'Turnkey' is an IT term used to describe a computer system that has been customized for a particular application. The term implies that the system has all the necessary hardware and software already in place but that the end user needs to turn the key to make it work.

A company's IT director or Chief Information Officer faces tremendous pressure to mitigate the costs and security hazards of mobility. This has been the catalyst for the development of systems known as Enterprise Mobility Management that are designed to minimize reliance on IT labor yet can support all the various networks and devices deployed by the enterprise.

The Cradle to Grave Approach

In 2011, Aberdeen Group, a research firm published its report on the state of enterprise mobility in the market and concluded that firms that outdo others are those that take a cradle to grave approach to managing the full mobile lifecycle. The study's main findings stated that enterprise mobility is past its infant state and has matured into a key business priority. Additionally, firms.... seeking to maintain or gain a competitive edge in an increasingly aggressive business environment explicitly state that... managing and securing their mobile devices and networks as their prime IT goals for the year Is this year correct?

The cradle to grave approach, which is the distinguishing mark of these market leading firms, aims at minimizing efficiency losses and optimizing performance along each step of the mobile ecosystem. Businesses know the future of their success depends largely upon their acceptance of mobility enterprise systems and look to them as the Holy Grail of the business world.

Andrew Borg, a senior research analyst at Aberdeen Group, has stressed that mobility is no longer merely a peripheral aspect of

IT- it is the new IT. Therefore, all mobility initiatives must now be merged and brought under the process-centric watch of an enterprise's central IT department.

Another important finding of the study is its estimation of the costs, or a regulation or compliance lapse resulting from the loss or theft of a mobile device. This loss can range anywhere from $10,000 to $491,000 per device: a colossal loss by all accounts that must be avoided at all costs.

It is naïve at best and criminally negligent at worst to assume that mobile devices such as smart phones and tablets that carry confidential corporate data will never be lost or stolen. The best an enterprise can do is to prepare itself for such an event by having the requisite enterprise mobility Management software in place, well in advance. Good software will allow users to wipe and lock the device in case of theft, locate the device if it is misplaced and recover a forgotten password under the watchful supervision of the system administrator. After all, wisdom lies in protecting what is precious and enterprise owners would agree that nothing is more valuable to the enterprise than corporate data.

Chapter 2

Popular Enterprise Mobility Trends

Enterprise mobility is a recent development undertaken by various firms and organizations. The introduction of enterprise mobility to the workplace means that employees can bring their own communication devices, such as laptops, notebooks, tablets and even cellphones, to work from. They also make use of cloud services, which various providers have recently started offering.

However, enterprise mobility is not solely limited to employees being able to work on their own mobile communication devices. In fact, it also allows the mobilization of corporate data. For example, if an employee is asked to go and visit a client to deliver a presentation, he can make the presentation on his own mobile communication device. He can then upload the presentation onto the cloud storage service his firm makes use of. When he reaches the client's meeting room, he can access the presentation and deliver it to the client.

Increased Connectivity

A growing need for increased connectivity is perhaps a major factor behind the growth in the popularity of enterprise mobility. Today, workers of the corporate world are constantly on the go. For example, certain position holders in an office, such as executives and managers, might be flying out frequently to other meeting points across the globe. In order to ensure access to all the files held by the firm, containing relevant information, and to ensure that the absence of an employee from the workplace or office does not hinder the completion of assigned work, there has been an increase in the demand for a method, by which employees of a firm can enjoy increased connectivity.

This increased connectivity has only been made possible because of enterprise mobility. Initially, most companies and firms chose to hand out their own approved devices to their employees. However, in more recent times, most employees have shifted towards using their own mobile communication devices. This may be because their own personal devices are more advanced gadgets or because they find it easier to use them.

There has also been a growth in the trend and the number of people seeking part time work, or those performing a large proportion of their tasks at home. For example, many firms now offer certain jobs that do not require employees to come into the office every day to work. Instead, employees are assigned tasks for a couple of days, such as writing reports or making presentations. They are then called into the office a few times a week for matters. And in some cases, an employee is completely virtual—working from home one hundred percent of the time.

Employees can now complete their assigned tasks from anywhere in the world using their own mobile devices and upload it on a cloud storage system. This can normally be accessed by the employee himself or by other employees of the firm whenever required. Moreover, enterprise mobility also allows employees to be constantly up to date with the company's agenda and calendar. Larger companies have also developed instant messaging applications that employees can install on their own mobile devices. This makes for an effective way of communicating with other employees and coworkers.

Technological Evolution

The technological evolution...has led to significant changes in what technology is being used for today. While pagers were commonly used thirty years ago, they have almost become obsolete and mobile phones have taken their place. Digital networks worldwide also support the new mobile devices that are available in the market today.

The evolution of mobility can be broken down into four major segments. The first phase was characterized by basic connectivity. This means that mobile devices were mainly used only for the call feature. Most mobile sets were large in size and did not support a lot of features.

In the second phase, connectivity expanded. There were improvements in battery life and technological developments also meant that manufacturers of mobile devices were able to build more compact devices that were easier to carry around than those that were used in the first phase of mobility. It was also during this phase that the very popular feature of text messaging was first developed and used.

In the third phase, sometime around the late 1990s, there was unwired enterprise. During this phase, there was a demand for greater accessibility. Networks evolved from 2G networks to 3G networks and many new mobile devices were built that also boasted the features of personal digital assistants, such as iPhones and Blackberries. With the emergence of such phones, users were able to remain connected with the workplace from wherever they were located in the phone, and could make use of important features, such as email and web browsing.

The fourth phase of mobility is that of automated interactions. This essentially means that one should expect to see certain automated devices and machines that are able to communicate without the intervention of any human. This is expected to fundamentally change the way business is carried out and managed by firms and organizations all over the world. For example, if a retailer is running out of a particular product, his computer might automatically send this information to that of the seller and initiate a transaction.

A Different Consumer and Workforce

The emergence of new kinds of technology has led to a different consumer and workforce. There has been an increase in the size of the consumer base and the workforce. Although the convenience of mobility was offered to very few workers initially, an increasing number of workers now expect the same kind of service. This is due to the fact that the trend of mobile business applications has increased over time. With a greater degree of mobility, more and more people are now willing to join the workforce as the option of working from home also opens up to many. Moreover, they do not really need to alter their routine if they wish to work; using mobile devices and mobile business applications makes it easy for them to access a firm's information and database from any place and at any time.

The increase in the use of mobile applications has also created a new consumer called the individual liable user. These consumers purchase their own smartphones, gradually wish to use the same kind of applications and have access to the same kind of content as corporations do. They are able to enjoy several different kinds of features, such as wireless internet and web based email and as their demand for more features and applications grows, providers will be seeking out new ways and methods with which to provide what the consumer is demanding.

Joining the Mobility Revolution

The mobility revolution has taken many businesses by storm. More and more people now prefer enterprise mobility because of the many advantages it offers to employees and the way it ensures continuity in work and prevents hindrances and disruptions.

However, it is not always easy to move towards mobility of the enterprise. There are several challenges that can be faced while undergoing a mobility revolution some of the possible challenges include. The sheer number of businesses switching to mobile enterprise systems requires service providers to be able to keep up with the demand.

- Technological development means that a wider range of mobile devices and software are now available. This means that providers of mobility services have to design services which can be used and supported by all kinds of software and mobile devices. This also means that providers are faced with the responsibility of developing policy controls and ensuring adequate security. This will ensure that enterprise mobility is cost effective and feasible.

- For enterprise mobility, different kinds of mobile applications are developed. However, this proves to be particularly troublesome for certain firms. Developing such applications and making them available to all employees is a complicated and tiresome process. Experts are required to devote a substantial amount of time to develop these mobile applications and ensure that can be accessed and used by all employees. Moreover, a substantial amount of resources is also utilized in the process, which may prove to be particularly costly for a firm.

- The help of experts and partners to guide them in technical areas and implementation of strategies.

Without this help, it is often difficult for firms to maintain a healthy rate of progress and minimize costs at the same time.

The City of Stockholm, Sweden

- **What is the City of Stockholm?**

  Stockholm is a sprawling city boasting a population of more than eight hundred and fifty thousand residents. Out of these residents, the city of Stockholm has around thousand workers working in sixteen various fields' social services, housing, education, child care, environmental protection, elderly care and more.

- **The Mobility Revolution**

  In order to provide information technology services to its employees, approximately ninety-five percent of the services is outsourced by the organization. For this very purpose, the organization has trained personnel who are involved in managing and overseeing the mobility contracts.

  Enterprise mobility has progressed leaps and bounds since 2006. Initially, around four hundred workers were provided with mobile communication devices that allowed them to access their email on the go and maintain an easy way of communication with their respective departments. At that time, all such devices were provided by the organization itself. In order to maintain strong security, the organization made use of a strict policy with regards to hardware encryption and data encryption and should this be marked as a registered trademark? Software was used.

  However, over the next four years there was an increase in the demand for email access on the go; it was no longer seen as an added benefit and it was now seen as a necessity. Therefore, there was an increase in enterprise mobility as even greater numbers of employees were now provided with mobile devices. The increase in enterprise mobility was also aided by the emergence of various new kinds of mobile devices and more advanced software, which supported specially designed apps aimed at enterprise mobility.

Despite the ease with which mobility services could now be provided to a to a greater number of employees, the existence of a wider variety of mobile devices and different kinds of software also introduced a problem in terms of security and management. As confidential information and files could be accessed by employees through a few clicks on their mobile devices, there was a need to ensure that proper security measures were taken to prevent data from falling into the wrong hands. Businesses had to expend their resources and energy to ensure effective management and proper security. Moreover, it was necessary to train employees to use the software properly. Use the mobility software.

In order to make the enterprise revolution easier, the city of Stockholm entered into a partnership with SYS Team. Together with SYS Team, the city of Stockholm was able to support enterprise mobility on a wider range of software and mobile devices. SYS Team also aided the city of Stockholm in expanding from around one thousand workers to more than five thousand workers. Of course, this also meant that there were a greater number of workers who had to be provided with mobility services.

Although the organization initially relied on a manual method for setting up mobility services for its workers, it presently enjoys the convenience of using an automatic process. Through the help of SYS Team, the city of

Stockholm can now make use of efficient and easy methods for providing, securing and managing mobility services on several different kinds of mobile devices. In fact, the city of Stockholm is hopeful they will soon develop...more than thirty applications for mobility services for.

Natuurmonumenten, the Netherlands

- **What is Natuurmonumenten?**
  Natuurmonumenten is a not-for-profit company based in the Netherlands that focuses primarily on saving and preserving the environment. The organization works... to protect animals and their habitats, plants, excursion areas, footpaths and more. In order to oversee the sprawling areas of over one hundred thousand hectares, Natuurmonumenten has more than seven hundred and fifty thousand members; six hundred employees and around three thousand volunteer workers.

- **The Mobility Revolution**
  Following a mobility revolution, the Information and Communication department of Natuurmonumenten aims to maintain connectivity through more than one hundred and areas across the Netherlands: providing services to more than sixteen hundred users and nine hundred personal computers. Additionally, one hundred and forty employees are field workers; workers to whom mobility enterprise is crucial to their job.

  The Information and Communication department of Natuurmonumenten has managed to allow half of these workers to access their email accounts via mobile devices. The rest of the workers are also made capable of accessing the company's inventory and information archives through their mobile devices. The company recently changed its mobile devices, due to the development of a new kind of application that was more suited to their needs.

Furthermore, the last year and a half saw many Natuurmonumenten employees using technologically advanced mobile devices. Many of them began bringing in these mobile devices to the workplace. The management of the organization took note of that. Together with the help of a consultancy firm specializing in providing mobility services, Natuurmonumenten developed strategies to support and provide mobility services to many different kinds of mobile devices. The consultancy firm also designed a roadmap in consultation with the management of Natuurmonumenten with regards to the course of action the firm must follow to have a successful mobility revolution. This roadmap: requires employees to use their mobility devices on the job. The company expects to implement and execute this policy over a time span of approximately six months, in the hope the mobility of enterprise will help all six hundred and eighty employees by improving performance and increasing their productivity.

Of course, as with other firms, an increased mobilization of enterprise also brings with it the danger of unauthorized personnel accessing important records and data. To curb this fear, the consultancy firm also helped the company devise and execute strict policies designed to maintain strong security and allowing the management to efficiently manage and oversee the employees.

- **What Did Natuurmonumenten Learn?**
  During its mobility revolution, the company realized the importance of profiling and categorizing all employees according to their mobility requirements. For example,

there were normal phone users who only made use of the calling service, light data users who would occasionally wish to send and receive email messages on their mobile devices, and heavy date users who were in constant need of access to their email accounts and access to the company's database. Once employees were profiled, policies were designed and executed accordingly.

The management of Natuurmonumenten also used the same mobility service to educate all of its employees on how to efficiently use the mobility services and maintain proper ethical standards at all times. Therefore, users of the mobility service were reminded of the responsibility given to them in ensuring that the service runs smoothly.

Norton Rose, United Kingdom and International

- **What is Norton Rose?**

Norton Rose is an international firm that provides services in the area of business law. It currently has more than thirty-nine offices located around the world in Europe, the Middle East and the Asia Pacific region.

The Mobility Revolution

Deciding to join the mobility revolution was an important turning point for Norton Rose. The mobility revolution meant that the firm's resources, database and experience could now be accessed by all its employees, regardless of where they are based in the world. Extensive efforts have been taken by Norton Rose to help in mobilizing the enterprise and ensuring that each lawyer has access to equal assistance and resources from the firm regardless of where or she is positioned. In this way, Norton Rose allows its employees to replicate the work environment wherever they are their main advantage in offering employees the mobility of enterprise lies in the fact that lawyers are no longer constrained with office hours and location. This is significantly advantageous to lawyers who prefer working on their clients' cases on site with the respective clients. The lawyers in this company also have the freedom to work during travel without interruption thanks to mobile enterprise. Furthermore, the ease with which lawyers can access the resources of Norton Rose from wherever they may be situated at that time also encourages lawyers to accept cases in other regions and countries.

Until recently, the information technology department at Norton Rose relied primarily on BlackBerry mobile devices and laptops for mobility services. However, new Smartphones and tablets came to be widely used by its employees, who then requested the information technology department to support all these various devices. Currently, Norton Rose's experts are working to support all devices used by employees. This will also help the company introduce a policy that will allow all of the company's employees to bring in their own devices to work. However, before formally announcing this policy, careful plans must be drawn up by the management to ensure that this policy will not compromise on the level of security the Norton Rose network currently enjoys. Moreover, Norton Rose should also aim design and execute the policy in in such a way that it stills allows the management to efficiently supervise the employees and the company overall.

At present, Norton Rose has entered into a partnership with Mobile Iron. In doing so, Norton Rose has achieved…an effective, superior strategy of bringing mobile access to all employees of the firm. At the same time, the strategy has been structured in such a manner that the firm does not run into problems in terms of security and management, and is using a single console to manage and operate the devices owned by the company itself and those owned by employees.

- **What Did Norton Rose Learn?**
Norton Rose has also realized the importance of localizing a base platform. This is important because it allows workers all over the world to unite because of a similar mobility service. To this effect, Norton Rose has developed a basic policy that is altered and localized across regions. Furthermore, sufficient progress has been made by the information technology department at Norton Rose; they are now aiming to develop a store for enterprise mobility applications, in accordance with the road map they constructed, which are expected to be supported by all the various kinds of devices used by the employees of the company.

Norton Rose is…careful and particular about the way it is progressing with the mobility revolution. Although the company itself also has a department dedicated to information and communication technology that employs highly trained staff, the company also seeks out the help and services of Escolar, which is an internet security consultancy firm.

Swiss Reinsurance, Switzerland and International

- **What is Swiss Reinsurance?**
As the name suggests, Swiss Reinsurance is a large and highly diversified reinsurance company. At present, the company operates all across the world in approximately twenty countries. The company handles several types of insurance including property insurance, casualty insurance and health and life insurance.

- **The Mobility Revolution**

The mobility revolution began for Swiss Reinsurance at the dawn of the twenty first century. From 2000 to 2010, all employees of the company were provided with Research in Motion Blackberry devices that allowed each employee to access his email account. Every measure was taken to ensure that the company did not face any issue with management, control and security. However, as also discussed in the case of Norton Rose in the United Kingdom, there was a growth in the trend of the use of iPhones. As a result, many employees of Swiss Reinsurance lodged requests with the relevant company officials to allow them to access their email accounts from their own personal mobile communication devices. During the year 2010, Swiss Reinsurance formally announced its first policy that allowed workers to bring in their own devices to the work place. To set up mobility services on their personal laptops and mobile devices, all the employees had to do was sign up with the relevant members of the information and communication technology department. There were two main reasons why the management at Swiss Reinsurance decided to opt for this policy.

Firstly, it obviously allowed for a greater degree of accessibility, flexibility and mobility, which, in turn, improved workers' efficiency and productivity.

Secondly, it also meant that a more democratic environment could be established in the workplace. If the firm carried on with its policy of providing company owned and company approved devices to its employees, only a selected set of employees were lucky enough to be able to enjoy enterprise mobility and the vast majority was still devoid of such facilities.

However, allowing employees to bring their own mobile devices to the work place meant that services could now be provided to whoever brought in his or her own mobile device and registered with the information and communication technology department. Furthermore, a bring your own device (BYOD) policy also meant that the information and communication technology department would be supporting more devices, different kinds of software and various mobile applications designed to aid the mobilization of enterprise.

The Bring Your Own Device (BYOD) program was formally launched in 2011. Initially, the BYOD program did not support Android devices because of a threat to security. Only Apple iOS mobile devices were allowed, as they supported encryption. Employees could access their own email accounts from their mobile devices. Other installed applications were also supported but they were constantly monitored by the information and communication technology department. As part of the security measures involved with the program, a set of guidelines was made available to all employees, which clearly stated how the employees were to use the mobility services, what was allowed and what was strictly forbidden.

After approximately four months of launching bring your own device program, the company boasts very promising statistics that reflect the popularity of the program amongst the company's employees. The main advisor and partner for Swiss Reinsurance during this mobility revolution was Nemesis. In Switzerland, approximately ten different devices were being registered each week for the program. Out of the devices being registered, there were approximately forty percent iPads and sixty percent iPhones.

- **What Did Swiss Reinsurance Learn?**

The mobility revolution taught experts at Swiss Reinsurance several important lessons. Firstly, they were fully aware of the importance of issuing a set of guidelines to employees who had signed up for the program. This is because they let the users know the ethics that were expected of them by Swiss Reinsurance: what was allowed and what was not allowed. It also told users that a certain degree of responsibility also lay with them and they would be held responsible in case of any kind of misuse of the program.

Secondly, members of the information and communication technology department were working to ensure adequate security to thwart any risks posed by bring your own device (BYOD) program. This is because Swiss Reinsurance was one of the first few companies to join the mobility revolution. As enterprise mobility was still a new concept at that time, there were very few firms that had progressed enough to consider bring your own device program. This meant that there were very few firms providing support in the way of security and management. However, while working to develop its own security policies, the information and communication technology department realized that it was a difficult and risky task to undertake solely by itself. Therefore, Swiss Reinsurance took on the services of a partner for the purpose of...designing and executing proper policies, such as those for detecting any kind of jailbreak and certificate delivery, and also implement policy as needed at the OS level.

Thirdly, the company also realized that it needed to come up with a proper roadmap and follow it through to successfully undergo the mobility resolution. As technological advances take place and new devices are made available on the market onto which the latest software has been installed, the need to support those devices and software soon arises. Therefore, it is important for the staff of the information and communication department to be fully trained so that they are capable of keeping up with the rate at which technological advances are taking place.

Important Lessons for Firms Undergoing a Mobility Revolution

After discussing the way these four different firms in different regions dealt with the mobility revolution, one can pick out certain factors and lessons that other firms should keep in mind when undergoing a mobility revolution, in order to get successful results in the end.

- Before a firm or company starts with any kind of mobility revolution, it is extremely important for it to have a properly developed and well thought out strategy to implement the mobility of the enterprise. At the same time, experts at the firm must also pay close attention to other associated policies, such as those dealing with the responsibility of the employees using mobility services, security and efficient management. Only once this has been done can a firm consider itself in a position to be able to chalk out a roadmap that it should try to adhere to. In order to be on the safe side, all firms and organizations are advised to devise a strategy that will allow them to determine where the company will stand with respect to mobility services within the next two to three years.

- In order for a firm to be able to deal with a mobility revolution, it is imperative that all employees of the firm must be well versed in the policies governing this mobilization and must be trained how to properly use the mobility service. When devising policies about security, experts should not only focus on ensuring that no unauthorized individual accesses the firm's exclusive archives and database. Instead, experts should also try to ensure that the policy is one that allows all employees to make good use of mobile services and explore the many features it has to offer while maintaining a safe and secure environment in the workspace. It is important for the management of a firm or an organization to realize that proper security is not only maintained by using advanced technology and well thought out policies, but that it is also heavily dependent on the way employees make use of the service being offered. In other words, training in regards to employee sensitivity and responsibility is essential.

- Apart from employee education and training, it is, of course, important that the policies are effective from both a management and security standpoint. This means that the mobile life cycle must be properly managed and controlled and special attention must be given to each of the following: device management, procurement, expenses, helpdesks, logistics and mobile app management. : One example of the necessity of careful management employee turnover. When someone leaves a business, he or she should immediately lose all access to both the reserves and the database on their personal mobile device.

- No one is unaware of the fast pace at which technological developments are taking place in the world today and the rate at which they are incorporated into the everyday life of their employees. Similarly, mobility also develops at the same rate. However, it is not always possible for the members of the information and communication technology department to stay current. They may be working overtime keeping up with policies and applications; making sure they are supported by the continual stream of new devices and software.

- Another reason for their lack of ability to keep up might be the fact that they are not trained and experienced enough to stay current with all the technological advancements taking place: ensuring proper security and reducing risk as far as possible and meeting the requirements of the employees of their company. Therefore, in order to ensure that the company is moving in the right direction with a minimal amount of risk Businesses must enlist the services of experts who have the most up-to-date knowledge and training available.

- Before a firm decides to be a part of the mobility revolution, it is imperative that management enlists the help of experts. Together they should analyze what has and has not worked for other businesses in the past to avoid making the same errors. : From there they can decide what will work best for their business and what will put them ahead of the competition.

- Having a rough idea of potential problems and issues will also give the management more time to think of possible ways in which to resolve them. Therefore, should the firm encounter one of these issues, it will be able to recover from the setback in a much shorter period of time than if the problem was completely unexpected. Moreover, the management can also get a rough idea of the expected direction that the company will take following a mobility revolution.

Verticalization

Statistics reveal that a growing proportion of the working population can now proudly call themselves mobile users. In fact, it is estimated that by 2015 the population of mobile workers in the world will amount to a staggering thirteen hundred million. This will comprise more than percent of the total working population. More than three hundred million people in the world have access to the internet. It is estimated that billion devices will be connected to the internet by 2020. It is also estimated that sixty percent of the world's population owns a cell phone. Studies have also been carried out to show the growth in the trend of using mobile phones and devices to access the internet. With it is believed that the number of people accessing the internet through laptops and personal computers will dwindle down to an insignificant amount and the number of people accessing the internet via their mobile phones will largely overshadow and outnumber them.

This growth in the use of mobility in enterprises is not some random occurrence. The change is deliberate and...can be attributed to several different factors. For example, it is believed that mobility in enterprises has been encouraged because of consumerization and BYOD to work policies developed by several firms, the emergence of new devices and the fast pace of technological development, cloud storage and other cloud based mobile applications and of course, Verticalization.

There are different ways in which enterprise mobility can be used. Some firms adopt a policy that allows employees to bring their own devices to the work place. Statistics show that, at present, percent of workers are choosing their own mobile devices instead of using devices provided by the companies themselves. When employees bring their own devices to the office, the information and communication technology department is responsible for providing them with the necessary applications that allow for enterprise mobility. Application stores are developed and the information and communication technology department is also responsible for managing all devices. In other cases, companies wish to mobilize existing applications; mobile applications are developed which will allow employees to access enterprise data. In some cases, firms allow employees to bring in their own devices and then mobilize existing applications.

The advantage of shedding the tiers encourages firms to opt for enterprise mobility. Data access and business logic are combined into business and data services, and presentation is done on mobile devices.

Companies are also able to centralize all cloud connectivity via integration. They manage applications that interact with social networks. All applications and services are managed by a gateway, which then relays the services to mobile devices. Services are made available for different kinds of phones including iPhones, Windows phones and Android mobile devices.

Consumerization

Consumerization of information technology is a trend that has recently grown. Consumerization essentially means information technology departments encourage the incorporation of devices owned by the consumers and design focused applications, which are relatively simple to use.

In the past, most technological innovations were made available only to those occupying superior posts and managing complex projects. Gradually, as progress was made, this technology was passed down the ladder to large organizations, customers and small businesses. Consumer markets, are now proving to be the driving force behind innovations and advancements in the field of information technology. There are many applications and devices that are initially targeted to customers and may later be business-friendly. When designing focused applications for consumer use, there are several things a company has to keep in mind. For example, it has to consider the basic design of the application, social dimensions, the user experience, interactivity and how appealing the nature of the content is. A company needs to design an application in a way such that it meets the requirements of its various consumers who are using different kinds of devices. Moreover, a firm needs to realize that a consumer using a desktop personal computer will have a better attention span as compared to one who is using a mobile device. Therefore, applications must be designed accordingly.

- **What Has Encouraged Consumerization?**

Consumerization has largely been encouraged by the policy adopted by many firms which allows users to bring their own devices with them to work. For example, they may bring their personal laptops, mobile phones or tablets and work with them. There are several reasons a growing number of businesses encourage this policy. Primarily, if employees use their personal devices, the cost of providing mobile devices is non-existent. Secondly, Employee-owned devices offers greater flexibility. This, in turn, increases their productivity and efficiency decreases disruptions in productivity and work-flow.

Thirdly, if an employee is working on his own device, he will most likely be able to handle the device expertly and will not need to be trained how to use a device (insert semi-colon after 'device'; saving time that would otherwise be spent training employees in proper usage of new devices.

Consumerization is also encouraged by other factors, such as using cloud services and applications offered by third parties. For example, a firm involved in manufacturing and preparing products for the market will need to scan barcodes frequently. Previously, barcodes could only be scanned using certain scanners and the like. However, technological advancements mean that applications dedicated to scanning barcodes have now been developed. If this service is only required by a specific department, then the company can think about issuing smartphones to employees of that department, which are designed to support the application. However, if a large number of employees are expected to scan barcodes, it would perhaps not be very feasible for the firm to provide smartphones for each employee.

This is no longer necessary, however, since the application is available for personally-owned devices. This cost-saving advantage also allows interaction between co-workers with absolute ease.

- **What Are the Advantages of Consumerization?**

There are several advantages of consumerization. Firstly, consumerization and standards are interrelated; they cannot be separated from one another. Consumer markets have developed, grown and evolved due to global standards in a competitive environment. For example, unlike corporate information technology networks, there is overlap present between several different services and products, such as text messaging or instant messaging services, laptop computers, tablets, desktop personal computers and more. There is no reason for this interoperability and wide variety of products from which to choose from will not last into the future. Advancements in web services will allow firms and organizations to integrate these systems into their business structures. This process of integration may prove to be problematic for some private business owners.

Secondly, there is also a great deal of convergence today. Looking back, one realizes that the most efficient method of marketing a product used by firms was to design a consumer device made to support certain kinds of applications. However, since that time, the mindset of manufacturers has changed and technological advancements have taken place. Now most consumer devices serve several different purposes instead of supporting only a couple of applications which limits their purpose. One does not need to look very far to see how consumer devices are being used for an ever-increasing number or reasons.

For example, there are many personal digital assistants (PDA) that overlap with mobile phones while personal computers and video game machines have started sharing many of the same features. In fact, many people feel that the popularity of personal digital assistants might be short lived unless they start supporting cellular networks and call features. This is because a typical smartphone, such as an iPhone, Android phone or Blackberry, also serves as a personal digital assistant bearing features such as notes, reminders, alarms, contact information, software for downloading, reading and writing documents and more.

As a result, more and more people opt to buy a single smartphone which provides all the necessary features offered by a personal digital assistant in addition to regular phone services such as calling and text messaging this points to the fact that converging work and personal time is another major factor in consumerization in the world of technology. Thirdly, consumerization is particularly beneficial to firms in the way of training and educating the employees. Many consumer systems train employees: relieving the firms of this crucial job. In fact, consumer systems are designed in such a manner that many employees are self-taught. There are several important services and systems employees are able to learn and use efficiently outside the office. Two excellent examples of this are IM (instant messaging) and text messaging.

Because of a growth in consumerization and consumrized products and services, it is highly possible workers and employees are actually more experienced, trained and technologically savvy than a business' IT department. This makes it possible for new-hires to insert him or herself into a new position without having to be trained in the basics of technological know-how. And it is all thanks to consummation.

What Are the Different Types of Consumerized Products and Services?

There are several consumerized products and services that are proving to be particularly popular, with web based email taking the leading position. Many firms and organizations now provide their employees with email addresses including the company's domain name, and employees can access their email from anywhere via the internet. They are also listed in the company's address book. This has led to a significant reduction in costs, as compared to those incurred when using internal systems.

Another popular consumerized product is the facility of Wi-Fi. Many firms and organizations now provide Wi-Fi access to both employees and consumers. For example, a fast food outlet has installed Wi-Fi in several hundred outlets, providing access to consumers and also using it disseminate information to workers and deal with payments made using credit cards. Although some might say that providing Wi-Fi is too big of a risk, it opens up many opportunities in the retail business.

Storage facility and file transfer is also another product of consumerization. Many workers use the internet as a storage facility; documents, files and presentations are uploaded onto the internet and can be accessed from anywhere in the world. Providers of such services charge a minimal rate for storage facility, which can also be used to transfer files. Many applications have also been designed for online storage and file transfer, which can also be installed onto mobile phone devices. Another similar service is that of automatic backup. Many firms now make use of automatic backup services, which create a backup of all files and documents without any effort made on the part of the user. This can be recovered and accessed via the internet by anyone who is authorized to access it.

An increasing number of firms and organizations are also making use of social networking systems. Although some people argue against their use because of security purposes, there has been an exponential increase in the use of instant messaging services, as it allows for efficient communication amongst employees.

Personal mobile phones are also being incorporated into the work environment. For example, an organization paid its employees' mobile service providers an extra amount to get them all incremental time on their existing phone packages, which was to be used exclusively for business related purposes. A close analysis of this experiment showed that the company was able to save a significant amount by indulging in this activity, rather than providing enterprise wide plans for all members of the company.

The use of internet fax is also emerging in several firms and organizations. The internet is used to both send and receive documents via fax. Anything sent via fax is delivered to the email Inbox of that particular person in the form of an attachment. However, the use of internet fax is still not as common as it could be, with many firms and organizations still relying on facsimile systems because of the paper documents involved. This is yet another technological advance likely to change and evolve over time, as paper documents are eliminated from the process.

Although the trend of consumerization can be regarded as a worldwide trend, the pace at which these changes have taken place is not the same. For example, the United States is currently taking the lead with percent agreeing to consumerization, Germany has percent and Japan is following with percent Likewise, industries have taken the lead when it comes to technological advancement. Education takes the lead with around eighty percent consumerization, health care has percent consumerization and the business services industry has percent consumerization. Those industries that are less consumerized include the manufacturing industry (with percent consumerization), the government sector (with percent consumerization) and the utilities sector (with percent consumerization). The degree of consumerization is not heavily dependent on the size of the company. However, medium sized companies (that is, a company employing approximately fifteen hundred workers) has a larger degree of consumerization (approximately, the adoption rate is percent).

New Devices and Device Categories

With rapid advancements in technology, it comes as no surprise that each week sees the unveiling of new kinds of technological devices. The use of mobile phones was not as common twenty years ago as it is now end Nearly everyone today can enjoy the connectivity and convenience of a mobile phone, whereas twenty years ago, they were enjoyed only by the most wealthy and elite of society.

However, mobile phones are not the only kinds of mobile devices that have emerged over the past twenty years. Instead of desktop personal computers, laptops are all the rage today. Due to their compact size and weight, they are easy to carry and are preferred by workers who are constantly on the go. Moreover, almost all features possessed by a desktop personal computer are now offered by a laptop as well: making laptops a more preferred device. Technological advancements also mean users can enjoy high memory space and disk storage space on their laptops.

Even more recent is the development and emergence of tablets. There are several advantages tablets provide to working people end they are extremely portable; easily tucked away in a backpack, purse or briefcase. They are especially useful for those who spend a lot of time in work-related travel. Tablets allow these (and others) to work at times and in places they would not otherwise be able to, i.e. airplane, car, train, etc. Because tablets support internet connections and all that this entails, employees are able to maintain constant contact with their work and others, as well.

There has also been a shift from the use of devices to devices owned by the employees themselves. However, the problem they then face is that employees bring all kinds of devices as well as a variety of models or types of the same device. Therefore, a company must ensure that their mobility applications are compatible with all devices in order to make them equally accessible to all employees.

In order for the information and communication technology department to provide proper connectivity and internet access to all kinds of devices, it is advisable to converge the infrastructure for both wired LAN and wireless LAN. This will also help the information technology department to monitor which device and which employee is connected to the company's internet. Secondly, the information technology department must also develop a unified and easy to use end to end architecture that provides for all of the firm's needs, such as servers and wireless access points. It also helps the firm to develop policies needed for mass mobility. If this approach is not used, it is possible that enterprise mobility will not serve its purpose of improving productivity. In fact, it might even do the opposite, that is, hamper productivity and efficiency.

Emerging Technologies

Enterprise mobility brings with it a variety of challenges and trends, such as consumerization of information technology, the policy adopted by some firms and organizations that lets employees bring in their own devices to the workplace and the need for strict measures of security. Thankfully, there are a number of technologies being developed which addresses these challenges as painlessly as possible.

Although most firms initially used mobile device management (commonly referred to as MDM) to administer the use of devices and implement policies, this was more suitable when firms handed out approved mobile devices to their employees. However, now that most firms allow employees to choose the kind of device they use (which is usually their personal device), technological advancements are required which allows for the challenges this can present.

One such challenge that has been faced and successfully dealt with is the ability for business applications and data to be used on a personally owned device without restricting personal access to that same device.

There has also been a separation of the provision and management of applications and devices, and the compliance and security of these applications and devices. Mobile device management is targeted at the provision and management of applications and devices, focusing on how to manage and implement policies on the devices being used. However, as the trend of using business applications has grown over time, there is also the emergence of mobile application management technology (also known as MAM).

Mobile application management focuses on the provision and management of mobile applications only, as the name itself suggests. Most firms and organizations now make use of both mobile device management technologies, as well as mobile application management technology. This is due to the fact that it allows businesses to control and manage the devices being used by their employees, as well the business applications they have made available to their employees.

As discussed previously, technological advancements have resulted in new kinds of software and mobile devices. As employees choose to move on to new kinds of mobile devices and software, they are naturally incorporated into the firm as well. Therefore, there have been significant developments in the area of device compliance and device security. Extensive work has been carried out to find solutions to address various problems. For example, technology has been developed to ensure the security of the firm's underlying operating system and solutions have been presented that let the management of the firm manage the devices and the way they are being used to ensure they are in compliance of the rules and regulations of the corporation. Developments in technology also mean there are solutions to other kinds of problems, such as jailbreaks and looting.

Apart from the area of device compliance and device security, there has also been significant progress in the field of compliance and security of business applications and data. This is mainly because of the growth in the number of firms adopting the policy. It is also because firms wish to separate the devices being used for the applications being provided, to be able to ensure security of their information and database. Technological developments have allowed firms to be able to distinguish between personal data and business data. Firms are obviously only interested in monitoring business data. Once a firm is able to distinguish between personal data and business data, it can then design policies and implement them to secure business data and regulate business applications.

By doing so employees complete control over the way they use and manage the personal data and personal applications on their mobile devices, while the place of business can keep a strict check on the business data and business applications on their mobile devices. Different kinds of technological innovations are designed for the distinguishing between personal and business data and applications. This includes secure containerization, application wrapping and virtualization, to mention a few.

- **What is Secure Containerization?**

Secure containerization is sometimes referred to as sandboxing. Essentially, it allows the firm to containerize (that is, wrap) all business data and business applications and subject them to the required security mechanisms, access controls and policy controls. Containerizing business applications and business data means that personal applications and personal data are not affected at all. Secure containerization is very helpful in providing for various different kinds of problems a business can face. For example, a business can think about the kind of security controls it would like to have on a device also intended for personal use. There are different questions a business can ask itself, such as whether it wants a mechanism for authentication of users, policy controls, and restrictions about the way certain applications or data can be used and accessed, respectively.

Another feature some businesses prefer to use is that of selective locking and the ability to delete all business data and business applications from a mobile device also for one's personal use as soon as the employee leaves the organization. This ensures sensitive information and data are not accessible by unauthorized individuals.

There are different types of secure containerization and businesses should use containerization type(s) based upon their security needs. Examples of applications targeted for security containerization include email, browsing capability and content download from various websites.

A firm also needs to consider the technology perspective when choosing a certain Type of secure containerization. In-client containerization is a kind of secure containerization that ensures business applications and business data are accessed only via an in-client model or browser. This ensures business data and information are not downloaded and stored elsewhere on the device. However, this is not sufficient for secure email. In application wrapping, individual applications are containerized. This is also used for maintaining secure email. By using application wrapping, a firm can also choose to wrap and secure only those applications it believes need to be strictly secured and monitored. A firm can also opt for a complete containerization. In this case, all business applications and business data are strictly controlled and monitored in a container. This is mostly used by those firms who wish to regulate and secure all kinds of features, such as web browsing, email and more.

Another issue of primary importance to: businesses regarding the issue of containerization is how to allow business applications to be linked with enterprise resources, while ensuring that personal applications are not able to do the same.

Whenever a business is deciding to embrace a newly-emerging technology expected to address certain generic concerns for businesses making use of enterprise mobility, it must always analyze its own expectations and requirements as well. For example, it may very well be that the features one business sees as being in need of stringent security, another business may not need at all.

Firms also need to keep in mind their requirements, such as whether or not they want to be able to wipe off all data and information from the mobile devices of employees once they leave the company. Lastly, the kind of technology a business embraces as part of their mobility enterprise system depends upon the extent of their bring-your-own-device policy.

The Next Major IT Paradigm

Many people regard mobility as the next major information technology paradigm. And there are a number of factors to support this belief.

Firstly, it is believed mobility leads to increased productivity. Because of various mobile applications that can be installed onto mobile phones and devices, workers are not restricted to the workplace and can choose to work from wherever they wish. Companies making use of the (BYOD) policy provide an incentive for workers to work more efficiently on a device of their choice, which they feel more comfortable using. Moreover, the design of applications is also very important; the easier a mobile application is to use, the more likely a worker is to use it leading to greater efficiency and more productivity.

Secondly, enterprise mobility leads to greater accessibility. Mobile devices such as phones, a variety

Laptops and tables can be used all over the world. This means employees can access email, the company database and other work-related information from anywhere in the world.

• Different Kinds of Users

Mobility is enjoyed by a variety of users. First, let's consider the casual user. A casual user is one who normally makes use of business applications on his smartphone or mobile device when he is on the go. The main reason they seek mobility is to save time; filling every moment with a degree of being connected. The advantages to a casual user of enterprise mobility are 1) he can better utilize ERP solutions 2) he can access important information anytime, anywhere. For casual users such as these, it is important that mobile applications are designed in such a…way that there is no difficulty in learning to use them. Moreover, they should be able to process the commands of the user very quickly.

If any upgrades are available for a certain business application, users should be able to easily install the upgrade.

Next let's look at the professional user. A professional user is one who relies heavily on his mobile device to conduct business. Examples of professional users include engineers, warehouse workers and maintenance staff. Professional users look to enterprise mobility to provide them with a greater sense of professionalism, the ability to easily upload information to a central location and obtain any other information required for that location and a superior service level. The benefits to Businesses that employ professionals of enterprise mobility are 1) lower administration costs 2) enhanced integrity of business data 3) easier monitoring of output and progress and 4) quick and easy access of multiple databases to assist in making sound business decisions. Lastly, there is the transactional user, who requires access to the ERP suite from his mobile device. Transactional users use several different kinds of mobile 'devices' as laptops, ultra books and tablets. The benefit to transactional users of enterprise mobility is easy access to all information needed to undertake business tasks. Moreover, using their own tablet makes them much more efficient, as they are experienced in handling the mobile device. Faster processing speed is also a key advantage.

Entering the mobility revolution is not without difficulties for any business, though—especially when it comes to the issues of security and the cost of integrating mobile devices and applications to the back end system(s). Several Businesses also make use of work processes that are not adapted for mobile use. However, the advantages of mobility, greatly outnumber the disadvantages and insert semi-colon after 'disadvantages'

SECTION 2 –

Enterprise Mobility in the Workplace

CHAPTER 3

Enterprise Mobility Team and a Mobile Workforce

Employee Oriented Mobility Use

What does B2E mean?

Before we dive into the business value of B2E apps, perhaps we need to make sure we understand what B2E means. For those of you who are familiar with the term, the section can prove to be a refresher, but you may jump straight to the next section to find out what's in store for your business if you invest in enterprise mobility.

B2E stands for business-to-employee. It is an umbrella term that refers to the entire exchange of information between the firm and its employees over the internet or an intranet. This intra-firm information may come in the form of company newsletters, operation manuals, policies, benefits and terms of employment.

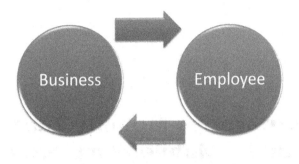

In the B2E or business to employee approach, the employee is at the center of the business unlike B2C (business to consumer) or B2B (business-to-business) approach where the consumer or other businesses are the focus respectively. The B2E approach spawned out of the persistent scarcity of information technology workers. Very generally, B2E encompasses all actions that a business takes to draw well-qualified staff towards itself as well as to incentivize it to stay with it in the face of fierce competition from other employers. These include but are not limited to aggressive hiring strategies and practices, benefits, opportunities for personal growth via education and training, bonuses and perks, flexible work hours and adoption of strategies that empower employees to emerge as leaders.

However, colloquially, B2E portals or people portals might also be referred to as simply B2E. A B2E portal is a customized homepage or desktop which may be unique to everyone in an organization. While the terms B2E portal and intranet are often used interchangeably, they fundamentally differ in their core focus. While the intranet's focus is the organization, the individual lies at the heart of the B2E portal. In addition to delivering tools typically found on an intranet such as sophisticated corporate directories, sales and customer relationship management tools and project management information, the B2E portal is more personalized since it also has the employee's personal information and web links that the employee might want such as stock updates or market trends. Some even have games to help employees unwind.

Over and above increasing the organization's efficiency, the B2E portal is aimed at increasing employee satisfaction and fostering a sense of community.

Organizations have the choice of developing their own B2E portals or acquiring the services of a professional B2E portal developer. Regardless of how they proceed, all B2E portals have the following features in common:

- It is accessed by a single URL by everyone in the organization.
- Some features are uniform to the entire organization while others are employee-defined.
- All B2E portals are flexible to the needs of a particular employee and can thus be personalized and altered almost effortlessly.

The Pros of B2E Apps

The past five or six years have witnessed an unmatched wave of consumerization of information technology and app usage. This has resulted in Enterprise Mobile Applications becoming far more user friendly than their preceding rather cumbersome form driven screens. While form driven interfaces try to resemble the traditional paper form, with the user pointing and clicking on options, choosing from drop down menus and filling in text boxes; the latest enterprise mobile apps have been designed to quite literally allow users to accomplish a wide Variety of tasks with the touch of a finger. This means business processes have become more streamlined and refined without the added cost of burdening employees with hefty user manuals and workshops.

App designers talk of a zero learning curve, meaning that the rate at which employees can become adept at learning to use these new apps proficiently is exceedingly high. This means that a considerable amount of effort and investment is put into designing the apps to make this possible. However, once in place, the enterprise mobile applications can enable businesses to reap many benefits, revolutionizing management, training and process conformity as well as improved throughput i.e. an increase in the system's capacity to deal with multiple people and issues within a particular time period.

Gauging how beneficial mobility is in terms of real business value compared to the traditional desktop or paper-based solutions involves making two critical assessments: is the use of the new mobile solution translating into more business for the firm? Is it leading to a cut in costs or an efficiency rise? To exemplify, think of a patient care system that allows round the clock monitoring in comparison to the conventional once-a-year appointment system. To successfully compete for funds for the installation of such a system, the stakeholders must be convinced of the real business value of such mobility-how many patients are benefiting from better care? How many potential emergencies are being avoided? If the benefit accrued exceeds the costs of putting a mobile system into place, only then can enterprise mobility be a viable solution for the organization.

Very broadly put, the pros of Enterprise Mobile Applications are twofold: the pros outweigh the cons and the quality is worth the cost.

Debatably, enterprise mobile applications ensure that the business always wins if put in place properly. This happens because traditional hindrances to businesses such as double handling, piling stacks of paperwork, travel, inventory holdings, mistakes and the time taken to uptake new resources and skills is reduced or eliminated.

Moreover, mobility use guarantees accurate information-- the timely provision of information. It also allows higher employee responsiveness; meaning employees can react to incoming data and instructions in a timelier manner. This, in turn, results in the ability for employees to get more done in less time. Employees also develop better organizational skills; planning, scheduling and coordinating their work activities with greater accuracy. By promoting process compliance, Employee Oriented Mobility solutions facilitate workers on the go and smooth communication between various subsidiaries of an organization. Today, the mobile apps in use in Nestlé's Switzerland office are the same as those in use by the organization in Pakistan; allowing for greater knowledge sharing. Finally, all these factors combined allow employees to spend more time on business activities instead of trivial IT administration tasks.

To get an idea of how your business can benefit from Employee Oriented Mobility Use, we look at specific domains of a business environment in the following section.

The Benefits of Employee Self Service

Gone are the days when employees needed to sign the daily attendance register. With an E2B solution in place, employees log their arrival and departure times into their timesheet directly via mobile devices. Day to day expenses can also recorded as they occur. In fact, receipts can be scanned and sent to the central system along with an expense claim in the blink of an eye. Moreover, activity logs can be updated on the go in real time instead of waiting until the end of the week or the month to do so.

As of mobility unnecessary trips to the office for many ordinary administrative tasks can be eliminated by working remotely. Information is more accurate and up-to-date; leading to better planning, organization and execution as well as saving both time and resources.

The Benefits of Customer Relationship Management and Sales

Today's tech-savvy consumers use barcode apps in their SMART PHONES to compare product prices to get the best possible bargain. And did you know a good number of consumers are paying for their purchases online via their phone while in the store itself rather than standing in a check-out line!

If they want to stay on top of t their game, the upwardly-mobile business will use this information and technology to boost sales and retain customer loyalty. The day has come for emailing coupons and discount codes directly to a customer's phone on a regular basis or (using GPS capabilities) when a customer in their database is in the vicinity.

E2B mobile apps have a number of benefits in the CRM and sales departments. Firstly, they allow employees to gather precise information about customers, activities and sales. Secondly, the clipboard can be chucked into the bin because thanks to mobility usage, there is no need to take notes and retype them later. Also, as described in the example above, employees can keep a close eye on customer leads and potential sale opportunities and act on them in Finally, not only do E2B apps equip CRM employees with tools for engaging with customers face to face, but they also free up time for the customer instead of pushing papers in the office.

The Benefits of Service and Plant/Asset Maintenance Sectors

With the power of E2B apps, an asset manager can inspect, update and create assets in his car, on his way to work or while running on a treadmill. Here too, the main advantage is information accuracy. The employee can keep himself abreast of the latest, most precise information regarding job transactions, assets and customers. Double handling is considerably reduced.

Organization skills improve. Employees can work from home or the work 'site' mobility usage also entails a sharp increase in the employees' speed to react to changes. Reduce the amount of time spent commuting and take payments—even providing customers/clients with invoices and receipts.

The Benefits of Tracking Stock and Taking Inventory

Imagine a vending machine "communicating" to a passing truck that its stock of candy is running low… sounds like a scene out of a sci-fi insert comma after sci-fi and change to read: doesn't it?? Think again. Mobile devices in the not-so-distant future may actually be capable of doing that and a lot more. With an E2B app installed, employees can capture and scan stock counts directly into a mobile device instead of lugging around bulky files from office to warehouse. These technologies promise to reduce the time spent on this essential, but time-consuming task. Paperwork can be done away with and all inventory details can literally be carried in one's pocket. Additionally, the number of errors is greatly reduced, taking inventory can be done more frequently and in a uniform fashion at multiple locations if necessary, adding new inventory is instantaneous and labels, tickets and reports can be printed almost instantly.

The Benefits of Maintaining Proof of Delivery

Courier services such as DHL and FedEx now allow consumers to track their parcels and provide them with recipient customer's signatures upon delivery. This has been made possible due to the increased reliance on E2B solutions. Employees can revise data on delivery, inventory and accounts receivable while on site.

In addition to the reduction in paperwork, inventory and information accuracy benefits accrued as in the case of other business areas, there is a marked improvement in the delivery performance of supply chains as measured by DIFOT (Delivery-in-full, On-time). Accounts receivable can be monitored more effectively. Errors and credit memoranda are reduced and customer signatures are captured. Lastly, cash reconciliation or the process of verifying the amount of cash present in the cash register at the close of business is made significantly simpler.

The Benefits of Training and Documentation

Valuable information regarding Occupational Health and Safety (OH&S) and Training can be communicated to Employees while they are in the field. Employee based Mobility Usage reduces traveling, ensures compliance and supplements OH&S. It can also be used to forward reminders and important documentation to employees on the fly

As explained in the above discussion, Employee-based Mobility Usage has far-reaching consequences for the whole business since it affects the efficiency of each part of the organization resulting in a compounding of benefits. Using these arguments as a starting point, IT departments in all kinds of firms can make a convincing case for greater investment in the development and acquisition of E2B Mobile Applications.

Five Benefits of Employee Mobility

Much of this text has been dedicated to the revolutionary idea of enterprise mobility and explaining how mobile is the new IT. Working remotely-whether from home or from the field- has been touted as an avant-garde business practice. But it is absolutely no use if it does not translate into real, tangible benefits for both the businesses and the employees involved.

With the advent of the SMART PHONE, the world has witnessed a radical change in the way people work. Soon, going to a high-rise building, sitting on your desk in a cubicle and returning home as the clock strikes 5 p.m. will become a Memory of a bygone era. Globalization has necessitated the growth of a mobile workforce and addressing this need has been made possible by the explosive growth in the development and use of mobile communication devices. Since all the data needed rests on the palm of their hands, employees in today's fast-paced era can be productive both inside and outside the office environment. To quote statistics, employees today work an average of eight more hours per week than they were before the mobility revolution; bringing increased productivity and efficiency to the business world.

How that has been made possible, you might ask. The answer lies in the greater freedom employees now enjoy when it comes to choosing their work location and time. Employees can work from home, consumer sites, public internet kiosks, subway stations to wherever. Not a single second need be wasted. Keeping these benefits in mind, more and more business are making enterprise mobility a part of their day to day operations by adopting approaches such as BYOD (Bring Your Own Device) and BYOA (Bring Your Own Apps) into the corporate environment. At the heart of it, a more mobile workforce has the following five main benefits:

Greater Flexibility

More employers today realize that each employee is unique in his or her working style, strengths and preferences. With that comes the realization that compared to a rigid schedule that ties everyone to the desk from nine to five, a more flexible working style enabled by enterprise mobility is very likely to optimize worker performance. Early risers can start working right after their morning jog without having to wait for the office premises to open. Night owls can get working on their assignments when they feel like it instead of lying in bed and counting sheep. Schedules can be customized to cater to each and every working style.

Reduced Absenteeism

Tackling employee absenteeism has always ranked high by companies globally. Absenteeism not only diminishes morale in an organization but also leads to low productivity and additional costs of hiring substitute staff. In fact, according to a Gallup survey quoted in Forbes Magazine, the lost productivity from employee absenteeism results in costs totaling to a whopping $84 billion!

Most employee absenteeism results from child or elderly care responsibilities, injuries, illnesses or bullying/harassment. These issues can be greatly reduced or possibly even become non-existent with the use of enterprise mobility. For example, an executive tending to a sick parent can at the minimum direct his subordinates while away from work hence finding a way around a potential bottleneck.

Increased Productivity

Technological advancements such as air cards, SMART PHONE, MI-Fi, Wi-Fi, tablets and iPhones mean the internet and cloud computing are readily accessible. Tasks such as editing proposals, conducting meetings, stocktaking and filing reports can be accomplished anytime, anywhere. Moreover, neither bad weather nor repairs or roadblocks can hamper employees from working.

Smaller Carbon Footprint

The elimination of unnecessary trips from the field to the office and vice versa can potentially lead to a dramatic drop in fossil fuel emissions. Think about it- you can do the environment a huge favor by submitting a proposal from your smart phone without driving all the way to your office. In addition, greenhouse gas emissions are curbed when the expanse of air conditioning is reduced since not every employee needs their own workplace.

Cuts brown Expenses

Since more of the workforce can work remotely, employers have to spend less on capital costs for offices and infrastructure. Employers are not the only ones who reap benefits of mobility as far as reduced expenses are concerned: employees can also save on transportation costs and parking charges.

There is no denying that that employee mobility has its advantages. However, a more honest depiction would describe employee mobility as a double-edged sword at best. With effective enterprise mobility management, employers that truly believe in the merits of a mobile workforce can make it a success. However, it cannot be stressed enough that a mobile workforce has its share of disadvantages such as potentially putting confidential and crucial data at risk. Therefore, if a company wants to utilize enterprise mobility as a means of increasing workforce mobility to gain a competitive edge in the market, it must spend considerably on the effective management and implementation of this tool as well.

A Quick Glance at the Disadvantages of Employee Mobility

To provide a balanced picture of employee mobility is essential to aid businesses in making an informed decision. Here, we summarize a few cons of a mobile workforce.

Over-connected

With internet connectivity, the lines between work and pleasure have been blurred which can result in the toppling of the delicate work-life balance. Most employees have an almost compulsive need to check their email every few minutes and fail to mentally disconnect from the workplace even when they are done with the day's work.

It may seem innocuous at first, but compromising on rest and your personal life can have negative repercussions on how you perform at work. Most employees express discontent at being chased by work even in the night or while they are away for a vacation. Employers also begin to harbor unfair expectations from their employees; expecting them to answer work-related mail and phone calls irrespective of their location or circumstances.

Not all alternate work locations are equal

We must remember that internet connections in public places can potentially expose private business information to all individuals using the public network. Moreover, while some workers may not be distracted by the cacophony in cafes and hotel lobbies, others might prefer a more secluded and familiar work area such as their own desk to perform their best. However, if the majority of team members decide on conducting a meeting at a coffee shop, there is little that a more introverted worker can do.

Detachment from company culture

Bumping into your coworkers at the water cooler or discussing ideas face to face in the meeting room fosters a sense of community and helps in establishing a strong company culture. Going to the office everyday allows employees to get to know each other and Creates feelings of belonging and loyalty towards the organization, which go missing when the workforce is mobile. Video conferencing may be a great tool for information exchange and knowledge sharing but the Creates experience and learning because of social interaction is considerably compromised by it.

When it comes to choosing workforce mobility or not, there can be no unambiguously right or wrong choice. Each business must assess the potential benefits of a mobile workforce against the possible pitfalls keeping its own goals and structure in mind.

Five Smart Ways to Spend on Enterprise Mobility

An old adage says you cannot make money without spending some. While the idea might seem obsolete in today's Creates business environment, where businesses are willing to go to any possible lengths to cut costs, things were not always as they are today. In the early 1980s, companies that made it big were those that were farsighted enough to envision the productivity rise and the competitive edge investing in desktop computers could give to their companies in the aggressive business environment. There is no deny in that money was money well spent.

Intriguingly, the same eagerness to spend on mobile technologies was absent when mobiles first entered the picture. Why do you think that was? For the most part, the problem was legacy. Businesses were slow to realize that the mobile was no longer nothing more than something to make a phone call from The business world was not quite ready or able to understand the potential mobile devices have to completely transform the way business

Today's CIOs are realizing that with smartphones, tablets and powerful B2E, B2B and B2C apps, IT has become synonymous with mobility. With employees having access to work 24/7 from any device they find convenient, the stereotypical 9 to 5 job is predicted to go the way of the dinosaurs-into extinction. In fact, the degree to which businesses embrace enterprise mobility and incorporate into their business activities has considerable predictive power in assessing which ones will rise above the crowd and beat their competitors. Another interesting observation is that smaller businesses, are more forthcoming when it comes to adopting new technologies.

Is spending on mobility wise? You bet! According to the projections from the International Data Corporation, by the year 2015, thirty-seven percent of the global workforce or 1.3 billion workers will identify as mobile workers. Capitalizing on this paradigm shift can only be possible if business grasp this opportunity to enhance their productivity by spending where they can get the greatest bang for the buck. Here are a few ideas on where to spend:

Think beyond smart phones

The fact that smartphones have increased productivity is a no-brainer. If you still want numbers to substantiate how much this productivity gain is in terms of real business value, the stats are as follows most businesses spend $120 per employee on mobility in a month translating into a daily expense of $4. This expense on mobility however enables workers to put in an extra eight hours of work on a weekly basis. Clearly, this is convincing evidence to demonstrate the impact that smart phones have had on productivity and the fact that the costs are outweighed by the benefits. But who says mobility spending has to stop with smart phones? To keep ahead of your competition, you can increase mobility even further by investing in tools like tablets, air cards and MI-Fi.

A heartening example of this is American Airlines' clever use of tablets to substitute bulky suitcases that accompany the pilots on flights. These suitcases weigh up to 35 lbs. and carry instructions pertaining to navigation. By compressing that data in a digital form and carrying it on a tablet, American Airlines saved on fuel costs and recovered the initial expense incurred by purchasing the tablets within days.

Think beyond mobile email

For the average businessperson, enterprise mobility entails a little more than checking your mail on the But for those who think out of the box, the possibilities are endless. It is worthwhile to spend time and money on researching and developing new apps as well as purchasing some of the already developed off-the-shelf apps that lead to greater productivity. Mobilizing customer relations management, field force and expense management using powerful E2B apps can enable you to stay ahead of the curve as your employees will be freed by the limitations of time and space at least when it comes to work.

Think beyond mobilizing just the sales reps and executives

The section "The impact of E2B on various business domains" above gives examples of how mobility can enhance the performance of the entire spectrum of business activities starting from sales to administration and asset management. While there might be a tendency to keep SMART PHONES restricted to just the sales reps and the executives, a little investment in pushing mobility down the organization's hierarchy can go a long way in enhancing productivity. When all employees can interact and make transactions in real time, there is bound to be a rise in the organization's efficiency and the flow of communication.

Why not mobilize things?

If technology continues to advance at the current rate, animals might not be the only ones on the planet communicating with one another. Confused? Perhaps you have come across the term the Internet of Things or more simply, machine to machine communication. There are many instances in the business environment where linking mobile devices to other computer systems and devices can reap immense benefits.

An example is the rental car business. Thieves sometimes rent cars for the purpose of stealing the car's newer tires and replacing them with older, worn-out tires. They then sell the new Tires and make a profit at the rental car company's expense. To get around this problem, CIOs have proposed installing devices on both the cars and tires that will alert the company as soon as the tires are removed Needless to say, the reduction in expenses resulting from this novel use of enterprise mobility can be substantial.

Similarly, to avoid unnecessary expense on roaming charges, your employees' mobile packages can be moderated remotely by synchronizing your corporate travel app with your mobile device management solution.

To recruit Millennial, start thinking like them

Most trends say that the 9 to 5 job might be on its way out, as more millennial join the workforce. If Generation Y has it its way, the traditional workplace might soon become a relic of the past. Workers today value their time and to them, a lucrative job offer is one that allows them to arrange their work around their life rather than having it the other way round. Therefore, even though a good pension plan is still relevant, companies wishing to employ and retain the best and brightest of the young lot know that offering MI-Fi needs to be a part of the package.

The recent, prolonged recession has led companies to become ultra-budget conscious. However, a timely realization of the immense transformative impact of technologies and the critical need to spend on them can make all the difference in today's competitive business environment In other words, today's workforce is as concerned about how they spend as they are how much they spend. Wise spending can go a very long way.

Empowering the Mobile Workforce

The past few years have witnessed such a rapid and unprecedented growth in mobile technology that just trying to keep up can often leave one feeling utterly lost. Many Business owners are struggling to understand the sheer expanse of the mobile industry-capitalizing on this ubiquitous connectivity is another matter altogether. But take heart and begin a new sentence. If you feel daunted, you are not alone and this book is meant to help you gain an understanding of what is out there.

With the advent of the SMART PHONE followed closely be the tablet and 4G make this phrase follow the comma, one of us has become mobile to some degree. While some have just dipped their toes into these unchartered waters and deployed basic email and mobile browser features, other business (especially startups) who wish to be more progressive, have taken to it like ducks to water and have embraced an entire spectrum of features and practices such as mobile CRM and inventory keeping, BYOD (Bring Your Own Device), BYOA (Bring Your Own Apps) etc.

The question facing managers new to the concept of mobility is where does one begin?

While the use of mobile applications to establish better communication with the business's clientele has been firmly in practice for some years now, facilitating interoffice communication between employees has not kept pace the same way.

Many experts agree that in order to attract and retain a young, Gen Y millennial workforce, it is important that businesses start thinking like them. That definitely entails greater connectivity so that workers have the flexibility of working from any location and can enjoy a flexible schedule. According to U.S. Census Bureau statistics, Gen Y millennial form forty seven percent of the population currently employed. In the context of enterprise mobility, this translates to almost half of today's workforce demanding greater access to information and interconnectivity on the fly.

Equipping every employee with the latest Smart phone or tablet is barely the beginning. For enterprises to truly benefit from mobility, the concept of mobility must seep into the very culture of the organization.

Begin by asking how each facet of the organization can potentially become more mobile. Do you see room for your managers to become more flexible with regards to the work environment of the employees that they supervise? Is the intranet currently in place being utilized to its full potential to facilitate communication within the office, to discuss ideas for new projects, to share information and documents? Such self-evaluation on your part will give you some idea of where your company stands when it comes to enterprise mobility, how far you might be lagging behind competitors and will provide you with a rough idea of what you need to in order to promote mobility in your organization.

Expecting a complete system overhaul instantaneously is unrealistic as well as impractical. It is best to roll out a mobility project in phases, starting with the departments that are better placed to act as catalysts for the transformation at all levels of the organization. Typically, the departments best-suited for initial mobility projects are those that make use of current data such as sales rather than those that are concerned with data aggregation and management such at the number crunching accounts department. This is not to say that the folks in accounts stand to gain nothing from the mobility revolution but rather that in relative terms, mobilizing the sales and inventory gives the greatest productivity boost to the firm if implemented correctly.

Section 2 gives some pointers on how E2B communication is changing the traditional ways departments such as employee self-service and inventory do business. Recent surveys undertaken by the VDC Research and the ADP Research Institute indicate that more and more companies are showing interest in utilizing mobile human resource solutions so that employees can enter Arrival and exit times into digital timesheets using their mobile devices, record daily expenses, scan receipts for bill reimbursements and payroll functions.

These ideas are intended to serve as a viable starting point in helping you empower your mobile workforce. As long as the mobile solutions implemented are easily accessible and user friendly without compromising on data security, they will certainly enhance your organization's productivity in part driven by improving employee satisfaction.

The following section presents a case study in this regard and will serve as an inspiration to businesses seeking to improve the mobility of their workforce.

Case Study: Amtel Mobile Device Management Secures Mobile Devices and Data Access for Bank Employees

The National Banks of Central Texas faced a dilemma that most organizations wanting to become mobile do- how do you make the right tradeoff between the flexibility and efficiency that come as a result of deploying mobile technology, with the hazard of data security breaches that are an undeniable feature of this connectivity? Bank data is especially confidential and if allowed to get into the wrong hands, can throw the whole system into utter chaos.

Initially, bank employees were issued bank-owned iPhones, iPads and Android mobile devices. Devices were chosen over devices to ensure stricter control over bank data. However as mobile device usage spread, the Banks felt that there was a pressing need to put additional regulations and controls into place. To implement this, they availed the services of Amtel Mobile Device Management.

The details of how the mobility solutions were implemented are beyond the scope of this case study but here is a summary of the end result.

- The mobile devices issued to the bank staff were secured.
- Configuration Profiles and Over the air (OTA) programming were utilized as security measures for disbursing mobile devices.
- A device tracking solution was installed in conjunction with a remote lock and wipe feature to mitigate the risk of data loss in the unforeseen events of mobile device loss or theft.
- The kinds of apps that could be downloaded on the provided mobile devices were restricted by clearly classifying permitted applications into an app white list and prohibited ones into a blacklist.
- The features available on smart phones and tablets were restricted to strictly those that are known to enhance productivity while those that diminish employee productivity were removed.
- Finally, call re-direction and real-time expense control measures were implanted that greatly reduced monthly bills and led to savings.

The Need for an Enterprise Mobility Team

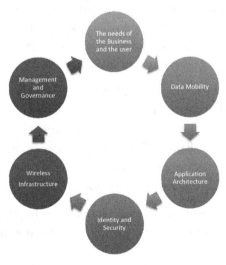

The above cycle tries to capture the stages, tasks and key players involved in an organization's approach to mobility.

While mobility has had a tremendous impact on the functioning of all departments of an organization, IT has undergone the greatest transformation. Before the advent of SMART phone shoes, tablets and mobile apps, IT enjoyed a supremacy over users in terms of the control and technical expertise it had over deciding which technological features were usable and feasible for the organization. With the arrival of mobility, however, this control has become a lot more diffused throughout the organization, which means that instead of an omnipotent IT department making all the decisions, there is now a need for several departments to formulate an Enterprise Mobility Team to work out what works best in the interests of the organization and its employees.

As the above figure tries to explain, the first step towards managing mobility is to do away with the legacy silo mentality. Silo mentality is defined as the attitude prevalent in certain organizations whereby departments and groups of individuals in the same company refuse to share information with each other for whatever reason. This compromises knowledge sharing and makes the company inefficient and on the road to failure. Contemporary management practices assert that in order to keep the workforce motivated and to establish a healthy working environment that stimulates growth, the silo mindset must be substituted with free flow of information across departments. This is important across the board but even more so when it comes to mobility management.

Organizations often suffer from the fallacious belief that disintegrating the organization into silos based on functions such as distinct and disconnected human resource, security, business management, legal, networking and application departments can improve efficiency. This ideology starts to crumble in the face of mobility problems since most of these problems are beyond the domain of a single department and in fact require input and collaboration from a number of divisions in the company.

Take for instance, the formulation of a policy on BYOD (Bring Your Own Device). As even a fleeting glance over the possible issues involved might reveal, designing a BYOD policy will not be possible unless input from all stakeholders is taken into consideration. This includes the users, business units, HR department, application and networking departments as well as the legal division.

Failure to acknowledge the need for a cohesive approach towards mobility can lead to the consideration a single aspect dominating the entire solution with other facets being completely ignored. For example in the case of BYOD Policy, security concerns might supersede other equally important issues such as quality of user experience and wireless LAN dependency. Mobility decision teams will inevitably run into dilemmas and will have to decide on the optimal tradeoff on a case-to-case basis.

To put it in other words, there is a tension of opposites involved. The five main dilemmas that an enterprise mobility team routinely confronts are the tradeoffs between network dependency, security risk, the cost of development and support, the quality of user experience and finally the cost of support and management.

The Business should aim for an architectural methodology when it comes to creating mobile solutions. With an architectural methodology in place, the IT department will have a structured approach towards approaching major mobility issues while giving due consideration to the interdependent nature the concerns of all departments such as business, legal, HR etc. An architectural methodology will also ensure that the right decision is reached when deciding between conflicting trade-offs such as cost of application development and security risk.

Key Players of the Team

To recapitulate, an enterprise mobility team is at its root, all the stakeholders in mobility issues working in tandem to settle conflicting trade-offs in a manner that is optimized for the organization's growth and future. It has a cross-functional structure, which is the polar opposite of the isolated silo unit models still in place in some workplaces.

The mobility team should work in tandem to devise a repeatable and justifiable methodology towards resolving interdependent, conflicting trade-offs. While the decisions made will vary from one case to the next, the methodology underlying them must not be random or arbitrary.

The key players that form composite units of the mobility team are:

- Eventual users
- Business management units
- Representatives from the human resource department
- Support personnel
- Application developers
- The Networking team which manages the hardware and software on the company's intranet, hardware, etc.

The mobility team is responsible for addressing all mobility issues with special attention to three fronts: the business, the user and IT. It needs to detail the stakes and each level and plan and develop solutions, doing justice to the needs of all the stakeholders involved.

The organization's mobility team must have the following objectives in mind:

- Tackle a wide range of issues that fall under the umbrella term 'mobility'. These include but are not limited to issues of data mobility, wireless connectivity, management and governance, application design and security risks.
- Make repetitious decision making easy by putting a well-structured methodology or work plan in place on how to resolve the tradeoff best when conflicts period.
- Keep abreast with the speed of the technology revolution and the business' evolving needs regarding mobility. The mobility team must have the creativity and the inquisitiveness to experiment with different possible solutions and to learn from experiences.

To put it in nonprofessional terms, there will be a constant tussle between various concerns such as cost, user experience and data security. To devise a solution that does not heavily favor one aspect of the debate while turning a blind eye towards the others, the organization needs a referee of sorts to negotiate between all parties and devise a solution all stakeholders agree upon. The mobility team plays that role.

What to Look for When Hiring People

The previous sections explained what a mobility team is expected to do within an organization. Regardless of whether the prospective employee is eventually going to represent the legal department or the networking department within the mobility team, two traits are of utmost importance: the ability to function as part of a group and mindfulness of the business and its clients evolving needs. Therefore, when hiring a mobility team member, an employer should look for the following: First and foremost, does the candidate have the skills and educational background to get the job done? Applicants often have a much more optimistic view of their capabilities than the real picture and as the 'you' are in a better position to evaluate whether there is sufficient evidence to support that they have what it takes to tackle the job. This does not mean that the individual must have years upon years of similar work experience but the candidate must be able to demonstrate the skill set the job calls for.

Since a mobility team will require extensive collaboration across departments, it is of utmost importance that the person you hire is reasonably easy to work with. Individuals that appear temperamental, pessimistic, or simply lacking in courtesy can demoralize the entire team and hamper cross-departmental communication. This is definitely a situation you want to avoid at all costs.

The mobility team will require working as a team, which is why it is important that all key players develop a good working rapport with one another. Communication and sharing of ideas across various segments of the team are vital for the optimal functioning of a mobility team.

That is why you will need to ascertain that the person you are hiring will stay with the company for a reasonably substantial time period--at least two years, but preferably for senior-level positions. Somebody who is looking for temporary employment or whose CV shows that they have a history of hopping from one job to the next will not be suitable for your organization's mobility team. The prospective employee must also have a demonstrable enthusiasm for the role they have applied for because job satisfaction can somewhat predict the employee's length of stay at the organization. Someone who is unhappy with what the job entails to begin with will not be productive and will become a liability for the team, dragging everyone's productivity down.

The prospective employee should be pleasant to be around. Somebody who has mood swings or is always throwing a tantrum will diminish the entire team's morale. This will impede the flow of ideas and knowledge.

Working on a mobility team is inherently stressful and will keep team members on their toes since technology is evolving at such a breakneck speed. Keeping these circumstances in mind, you will need someone who can put up with these negatives ideally someone who views these difficulties as challenges and relishes them. At the very least, the prospective employee must show some evidence of being tolerant and patient in the face of unforeseen problems and the resilience to not crack under stress.

Every company has its own unique culture. Some don't mind their employees showing up in casuals and referring to the boss by his first name, while others require more discipline and calling Mr. X 'Steve' can get you in a whole lot of trouble. The company's culture permeates every facet of the operational dynamics therefore, at the time of hiring; you must look for people who you think can potentially assimilate in the company. Their own personality and work ethics must not be too different from the kind of culture your company has otherwise both parties will struggle to come to terms with one another. The ideal situation is one where the new employee meshes into the prevalent culture with ease.

Look for someone with a strong work ethic. That means someone who isn't merely physically present on the office premises and procrastinates just to give the impression of being busy. Work on the mobility is interdependent, with the output from one team member acting as input for others. Someone who tweets all day or cannot get off Facebook is unlikely to get his share of the job done which can have dire repercussions for everyone involved.

Look for someone with a spark of passion rather than anyone who sees the position as just another job out of hundreds. The more enthusiastic the employee is about his field of work, the more productive he will be and the more the team can get done. Also, look for someone who has the openness of mind and a natural curiosity to devise out-of-the-box solutions for the many issues that might arise.

Finally, the job will require someone who does not become too comfortable in one state of affairs and can has the adaptability and flexibility to not only learn about how the new technologies function but also devise new methods of deploying those technologies to benefit the organization.

An employee that fulfills the above stated criterion can do wonders for the mobility team. Of course, the perfect candidate is tough to come by and certain compromises must be made at times but the above list will at least serve as a guide in deciding what to search for when hiring.

Taking the Help of Mobility Experts

With new technologies hitting the market every day, the way the world conducts business has changed for all times to come. Moreover, this revolution is showing no signs of slowing down. SMART PHONES, tablets, cloud computing and wireless internet have made connectivity ever-present and it has become critical for businesses to capitalize on the mobility that these technologies bring, so as to remain at par with competitors. Where does one begin? Planning, installing and managing such a system can seem to be a Herculean task at first; that is why you need to turn to experts.

Mobility management has become a specialized field. Most mobility experts offer a wide range of technologies tailored according to their clients' need and scale of operations. Not only do they have a repertoire of several solutions, these experts are also trained to implement them in the most cost effective way possible. As any seasoned businessperson would tell you, being able to spot sales opportunities while keeping costs to a minimum is a guarantor of making it big in the market.

Most experts offer a range of networking solutions and are educated in how these technologies work as well as how they can be combined to benefit the client most. With the help of specialists, you can deploy a solution that is well supported, protected and built-to-last. The products offered include but are not limited to Wireless Local Area Networks, Video Conferencing Systems, Point of Sale Systems, Forklift Mount Terminals, Digital Signage, Handheld Computers, Mobile Devices, RFID (Radio Frequency Identification) Technologies, Networking Hardware, Self Service Kiosks, Audio / Video solutions, Mobile Application Development, Mobile Platform Development, Mobile Website Development and Mobile Porting Solutions.

Mobility experts also offer diverse services such as the development, implementation and management of deployment programs, project management, installation of wireless and networking solutions from scratch or bringing existing systems up-to-date, maintenance post installation. Most firms offer on-site as well as remote surveys employing innovative technology to evaluate the business's networking requirements.

Aside from getting the right hardware in place, it is of utmost importance that the key managers and other concerned personnel receive well-timed and comprehensive instruction on how to make the most of the installed equipment and deal with any glitches that might pop up. In the absence of such training, any equipment, however advanced or expensive, is as good as nothing at all. Obtaining the services of a good mobility expert organization will ensure that your firm optimally uses the acquired machinery and gets the intended business boost. A quality mobility expert firm will offer you a combination of

Mobile Technicians, Technical Support Staff working hand-in-hand with experienced Quality Control Personnel to impart the requisite training your employees need to operate the hardware with the confidence and skill of a pro.

The benefits your business can reap from investing in mobility are perhaps best illustrated by the case study in the following section.

Case Study: XTRA Lease goes mobile with Velocity Inc.

The purpose of this case study is to demonstrate that recruiting the help of a team of mobility experts can help companies devise the best possible mobility solutions keeping their structure, scale and requirements in view.

Velocity is a mobility specialist firm that serves a wide spectrum of markets ranging from education to government bodies to retail and distribution. Recently it extended its Wireless Local Area Network WLAN Services to XTRA Lease. The installation spanning 97 sites took 10 months to complete. The client, XTRA Lease, is a logistics firm with its head office located in St. Louis, Missouri. Basically, it rents and leases trailers for-hire and private motor carriers. It boasts a fleet of 125,000 trailers and other heavy machinery including vans, reefers (refrigerator trucks), flatbed trailers, storage trailers and specialty trailers.

When hiring Velocity to acquire mobility solutions, XTRA Lease had the following end in mind:

- Meet customer requirements across its rapidly expanding network.
- Obtain inventory statistics in real-time for the most efficient supply chain management.
- Trim down transaction costs while catering to a larger number of clients.

With these acting as guideposts, XTRA Lease collaborated with Velocity to form its mobility strategy. To improve the efficiency of routine core tasks at all its branches and the quality of service delivered to its customers, Velocity designed a mobility solution by the name of XTRA Xpress service incorporating feedback and suggestions from the key personnel at XTRA Leasing.

A customized application for tablets facilitated the speedy processing of customer transactions and trailer inventories on the go. Once designed, the task ahead was to install and deploy it at the least cost at XTRA Lease branches across North America. Another challenge was to seamlessly integrate the new app with XTRA Lease's already instated legacy system.

Velocity Inc. aided XTRA Lease in developing the requisite specifications for wireless hardware including the mounting of special wireless antennae at the optimum signal receiving locations at each facility. To get the entire 80-branch network mobile, the wireless 802.11b infrastructure was installed which also facilitated the timely rollout of the service network-wide.

Upon completion, XTRA Lease's Senior Vice President Steve Zaborowski was all praise for Velocity Inc.'s handling of the project. Hiring the expertise of mobility specialists for the project benefitted XTRA Lease in many ways; particularly worth mentioning is Velocity Inc.'s yielding, customer centric approach towards executing its solutions and its ability to expertly detect possible technical improvements that XTRA Lease's management might not have thought of to begin with.

In summary, Velocity Inc. installed quality wireless infrastructure throughout XTRA Lease's branch network at very economical rates compared to its competitors. Moreover, the installations were supplemented by in-depth reports accounting for the specifics for each facility.

Because of its many linkages (such as those with hardware suppliers and software developers), a mobility expert is in a stronger position to negotiate the cost of the project compared to a solo businessperson on his own. Moreover, years of experience across a diverse range of clients endow mobility experts with the foresight and business acumen necessary to surpass their client's expectations by identifying possible room for improvement where the original plan might be lacking. All these elements combined make for a mobility solution that is bound to take your business's productivity new heights.

CHAPTER 4

Enterprise Mobility Devices and Platforms

With global markets and growing business needs and requirements, it is necessary to address the need for enterprise mobility in all sectors of business activity. All enterprises, from fast moving consumer goods corporations (FMCGs) to industrial businesses, need to be on their toes, as activities span over continents and diverse markets. For increasing mobility among workers in the enterprise, the right devices and technology must be identified and chosen to be then distributed and implemented on different levels of the organizational structure.

Making Device Decisions

It is a thing of the past to assume that a company needs to have a multimillion dollar to be able to invest in high mobility decisions. Today, even, the smallest firms are engaged in outsourcing labor and raw material from other countries and subsequently selling their output across different continents. : For companies such as these costs are generally higher and profit margins low at the initial stages of activity. How do such firms ensure enterprise mobility as a core characteristic in operations? Investing in enterprise mobility software such as SAP would not be recommended for such companies as the returns are unlikely to pay off the extremely high initial investment. The high cost associated with the implementation of such resource planning and enterprise mobility software is the main reason why they are substituted by other devices, such as cell phones, laptops and tablet PCs.

Firms widely adopt several types of devices to be used in business activity. From high range laptops specially designed to perform well for the needs of the business, to BlackBerry cell phones that increase the availability of officials and real time responses to emails from all over the world, companies are investing greatly in technology to increase mobility of the workforce without having them leave the geographical boundaries of the country where they operate. The reason for highly personal devices being used in official sectors of business activity is that the features required by businesses that were previously available only in enterprise devices, are now widely available in consumer devices such as laptops, cellular phones and tablets. The fact associated with this is that the enterprise devices which were expensive to buy were bound to become outdated soon, and changing needs of the business required diversity and flexibility in the devices. All these factors can be conveniently found in the now popular Smartphones and special made to order laptop and tablet devices. Companies find it a better alternative to invest in these for achieving mobility needs.

A firm makes a device decision depends on several factors.

Budget Allocation

How much can the business allocate to the purchase of these devices? This will depend on the size of the firm (naturally a multinational corporation will have more funds than a small privately owned business to allocate to such devices) and the number of employees that the device will be bought for.

Organizational Structure

The needs of the business will have to be identified. Which levels of the hierarchy require devices for mobility will have to be decided upon. Surely, the earning assembly line workers have no need for BlackBerry cell phones, as they do not need to communicate with company officials or assess the production or sales flows. The levels of the organization where the device implementation will be made need to be decided.

Organizational Needs

The work requirements of the officials need to be identified. Which software, applications and devices are suitable for the work will be decided upon, once the needs of the officials and their work plan are identified.

Value for Money

Each business wishes to achieve cost efficiency. With the budget allocation that is decided, the device that can meet the most organizational needs will be the one that the company chooses to buy for its employees. This is to optimize the value they receive from the money spent on the device.

Return on Investment

It is good to estimate the lifespan of the device. How long is the investment intended to fuel business needs? If the estimated life of the device is 4 years, a cost benefit analysis must be carried out to ensure that the investment pays for itself and adds value to operations. An important factor to be considered here is the repair costs, the chances of damage occurring to the devices, and the cost of updates: required while the device is in use.

Given the above factors, a business can devise a device policy and choose the device that provides optimum facility and ease to the workers and the business in aggregate.

What should be noted is that BlackBerry phones, previously essential for enterprise mobility, are now hurriedly being replaced by the iPhone, Android and Windows phones. This is happening because they offer many different executive apps and options to carry out executive tasks on a small device which BlackBerry did not offer.

Top 5 Gadgets for the Workplace

Business organizations are heavily investing in IT and devices to help raise worker productivity and mobility. The pressing need to be able to communicate in real time with other entities such as consumers and other companies all over the world makes every company strive to find the best devices and gadgets they can put to use in the workplace. Such gadgets include all devices that help the company become more mobile and make services readily available to all stakeholders.

Tablet PCs

Possibly the most convenient innovation in the world of technology, tablet computers are the "in thing" among company executives nowadays. Determining which platform is the most popular in tablets is a no brainer, as most people prefer Apple's iOS, followed by an android, and the last in the race to sell in the executive market are Microsoft and BlackBerry. Apple and Android platforms are the market leaders in tablet sales, and are constantly in fierce competition with each other

Why Companies Choose Tablets

Tablet computers are easy to use, provide workers with access to internet (some can also access company intranets and VPNs) and easy to use applications that let them work when they're not physically present at the workplace, allowing them to use applications and software that can normally be found in any traditional computer. Many multinational corporations have rolled out thousands of tablets to the workers to reduce lag in working and ensure highest efficiency. The question of security is also there in tablet users but most companies ensure that their choice of the tablet has security measures and policies in line with theirs; for example, emails and data in iPad are encrypted. This ensures secrecy and privacy of data in the network.

BlackBerry Devices

The company specializing in producing phones and tablets to facilitate the corporate world, BlackBerry is still considered the best for many executives, and even government authorities. The crowning point is that people can access their email and other documents while travelling or from home. Essential business tasks can be easily performed on the BlackBerry device without any complications, enhancing enterprise mobility to the maximum. This remains the most popular platform for executives even now. BlackBerry has been more consistent with updating their cell phone devices rather than tablets, and this is where they excel.

With their new BlackBerry Enterprise Service (BES) introduced with BlackBerry 10, the option to use it for Android and Apple devices ensures the ultimate enterprise mobility, and BlackBerry can be used across different platforms. This has definitely given BlackBerry an edge through cross platform compatibility, and the services of android or iOS phones can be used through BES. And even though BES is a recent development, firms all over the world have been giving their employees BlackBerry handsets for many years as it makes one fundamental thing possible – enterprise mobility.

Android Phones – Samsung

Samsung has been releasing very cost effective cellular phone sets with a wide range of features for many years. The most recent series that Samsung introduced in the market was the Android Platform Galaxy series which entrepreneurs grabbed as soon as they were available in the market. Their newest release is a completely new security platform (combating BlackBerry BES) known as Samsung KNOX, which was aimed mainly for enterprise use. The fact that BYOD (Bring Your Own Device) is widely welcomed by firms means people investing in the Samsung Galaxy phone will have the opportunity to use it to increase their work efficiency at the office, and the existence of KNOX answers all security concerns that any company may have about a private phone being used for corporate data and details.

Pocket Projectors

Going slightly off from the previous gadgets in focus, a business not only needs devices to manage communications, orders and shipments, but also needs gadgets in the workplace to increase efficiency hand in hand with mobility. The introduction of pocket projectors in the corporate field is a great new investment, as it reduced the need for one workplace to invest in multiple projector rooms, most of which are left unused and unattended for the most part of business operations. The pocket projector is an efficient tool that can be hand carried from room to room, and connected to your laptop or cell phone device to project images on any screen. Decreasing the need for large office space and investment of thousands of dollars in large projectors can be cut, and mobility will be ensured as official meetings and presentations can be carried out anywhere, by any employee who is in possession of this helpful gadget.

Laptops

The most generic and well-known, laptops are a necessity for all workers in the corporate world. These portable computers come in many makes, shapes and sizes, each customized to serve the needs of the user. They are the most flexible and personalized enterprise mobility gadget, and used by millions of businessmen daily to carry out essential work related tasks and operations.

Increased Efficiency through Enterprise Mobility

There is no doubt efficiency is raised through adoption of enterprise mobility devices and platforms. In many facets of the firm's activity, enterprise mobility is enabled through use of devices such as those mentioned above. This is through many factors, both while in and out of the office.

Efficiency at the Workplace

It is equally suitable to employ mobility devices to raise efficiency for employees in the physical workplace. People usually have their cell phones synced with their email and other updates such as calendars and reminders. Using multiple devices to carry out a task is tiresome and wastes valuable time. For example, a worker will get a notification to clear an important check for a client though his Smartphone, but if he doesn't have access to the corporate account on his mobile, he will have to move this task to the computer provided to him by the company to complete the process. This adds steps to work that could simply be done if the worker's handheld device was integrated with the company and its working, and trivial to important tasks could be carried out quickly and with efficiency. By eliminating time consuming steps of the work procedure, efficiency will be raised while the employee is at work.

For companies whose operations span over vast geographical areas, travelling costs are estimated to be extremely high if officials are to travel to different areas for corporate meetings and decision making. It is essential, therefore, to be geographically mobile in order to increase productivity and decrease the stress and strain of business travel.

A way to make workers more mobile and keep them from experiencing the physical strain of travelling is to enable software and platforms to boost communication across geographical boundaries. The construction of virtual conference rooms can enable employees to attend meetings in different geographical locations without leaving the office building. Audio and video conferencing is an essential aspect that needs implementation to raise enterprise mobility. It will ensure that the worker does not have to give up valuable work hours to leave the city for business. It can simply be done through internet based services.

Traditional business practices make work monotonous and simple tasks appear to be mundane activities. Workers can easily lose incentive when they are made to work on computers and sit at desks for hours on end. Several workers may feel that their health is being compromised while they work this way. One common example of this is Carpal Tunnel Syndrome, which occurs when someone engages in too much typing for extended periods of time. : Additionally, back and neck problems may also be experienced by clerical workers who spend a lot of their time working at desks. Giving portable devices to employees makes them more versatile and mobile in their work, and they may feel less pressure of spending their entire time at a desk or a computer. Moreover, the simple and user friendly interface of mobile devices and tablets makes small tasks even easier to accomplish through the click of a button. This can make the work load seem less and put less stress on employees, thus making them more efficient in their working.

Giving employees access to the outside world, the enterprise can enable them to constantly be in contact with foreign operations. Shipments of raw materials can be tracked online to determine which point of delivery they have reached, and order dates can be finalized for customers accordingly, and any changes can be made immediately if some delay is expected. This access to essential details at all times can keep employees on their toes: regarding business operations.

The most important thing mobility devices do is reduce clutter at the workspace. Instead of a laptop, a PC and all the accessories that come it, workers can be introduced to more mobile and flexible devices that can be easily moved around at the workplace. The offices and workstations become less cluttered; giving employees a clean free space to work and brainstorm. This has been said to improve worker morale and productivity greatly.

The return on investment (ROI) for mobility devices is very high for companies and the increased worker productivity is seen in terms of higher returns to the company. Usually the investment pays for itself very early into the use of the devices, and the company keeps gaining over the period of use.

Efficiency outside the Workplace

Work efficiency is not increased only while the employee is in the office. In many ways, productivity is enhanced in indirect ways, even when the employee may be at home, travelling, or on holiday.

Through mobile devices, employees have access to corporate information wherever they are and can use it to deal with customers and suppliers when needed. This helps create strong bonds with customers and suppliers, thereby enhancing the company's image and creating an efficient work environment, even when the employees are not physically present at the workplace.

Customer Relationship Management (CRM) is an essential task for every enterprise and this is exactly what investment in mobility devices provides to a company. Mobility facilitates relationships with customers through the simple attribute of; providing real time access to customer queries and an option to respond to them as quickly as possible. If customers feel they are the top priority and that their needs are catered to whenever they demand, the company's CRM position is likely to strengthen. Workers anywhere in the world can help customers with their problems, questions and advice with mobility devices, without the company needing to invest heavily in specially designed CRM software (which is mostly immobile when installed in fixed enterprise devices.

With access to cloud computing on mobile devices, executives can make important business announcements and changes anytime...anywhere. Efficiency is never compromised, due to the fact that physical absence from the business does not affect one's capabilities to run their business from a mobile device.

The Best Apps for Increased Workplace Productivity

Using consumer devices in the enterprise is a common decision made by most companies nowadays. Gone are the days when extremely complex and technical systems were purchased and installed for each worker. This may be true for many enterprises, but less so for those seeking to achieve high mobility to raise performance. Companies invest in tablets and mobile phones to raise mobility and several platforms are used by the brands that make these devices. This section will focus on enterprise use of consumer products such as tablets and Smartphones, and the applications they provide to facilitate business activity.

Applications in tablets and smartphones (known popularly as simply "Apps") are available in all iOS, Android, Windows or any other smartphone provider. These apps, originally designed for consumers using the devices for personal use, can be found in practically every category imaginable. For example, for a college student, the Wolfram Alpha app is essential for any calculus assignment (Yes, it makes your assignments for you) and for anyone wishing to bake a cake, Betty Crocker's mobile app will teach you how. There is an app created for everything in the world ranging from guitar lessons on your phone to travel planner apps that help you plan your vacation.

In recent years, app developers have not only been concentrating on end consumers as the audience for their apps, but large quantities of apps have been specially developed to facilitate the workings of large enterprises. For example, Standard Chartered Bank, to which has operations all over the world, has adopted Apple as its official device provider for all employees. From iPhones to iPads, all workers are provided with these devices which are used to perform tasks on apps specially designed for Standard Chartered, providing utmost security and secrecy, as these apps are limited to Standard Chartered workers and customers only. Customer queries are answered through chat on Apple apps, and banking is done on mobile devices. The need for personal face to face communication has fallen and the enterprise uses these consumer devices to facilitate it's working.

Going into a broader spectrum, we will discuss the apps being used in the corporate world these days. These are available to be downloaded to be used on any smartphone. While some apps are free, others will require using your credit card or other bank information to purchase them.

The top apps for executives are as follows:

Cisco WebEx: Cisco is software used by many companies to hold extremely realistic virtual web meetings, which include executives from all over the world. Having an app for Cisco in order to participate in video conferences means that business executives can be a part of their business meetings regardless of where they are.

Evernote: This is an app for taking quick notes and highly useful for people who do not have the time to sit down and spend a lot of time and effort on preparing notes for future reference. Notes can be taken using this app in the form of pictures, text or even voice. Evernote can help employees keep data and make sure they do not miss any important information when on the Insert period.

Drop Box: For storing large amounts of data, is convenient for anyone wishing to use cloud computing. The use of Dropbox is commonly used by a variety of people. Its popularity stems from the fact that it is safe, secure and can be accessed from anywhere through an internet connection. Another plus point is that you do not need to manually sync your phone with your Dropbox. It is done automatically when a sound internet connection is available.

A similar app is that of Google Drive. Google Drive works similar to Dropbox and serves as a cloud store for data.

Omni Focus: This is one of the best planners for people who have many commitments to meet in a day and several meetings to attend. This paid app keeps track of all meetings relative to location, entity and time, and notifies the person when an important commitment is near.

Pa Safe Share: This app helps business executives feel comfortable when using a mobile device for official data. It ensures safety of corporate information.

Qantas Windows Phone 7 app: This app is especially helpful for travel. It is a travel assistant with the option of mobile and boarding card which you can get on your phone. It also It also updates the user on weather conditions of your destination, flight information for continued travel and other pertinent information.

Yelp: If you are in a city you are not familiar with, you will appreciate Yelp, Yelp, an iPhone app that helps generate customer reviews of nearby restaurants and hotels. It uses GPS to track your position and helps you decide where you need/want to go.

Cloud On: This is an app for iPhone and Android phones used by workers to edit, change and read files in Microsoft Excel, Word and all parts of Microsoft Office. It enables workers to create and edit files on their phones or tablets, and can retrieve files from cloud storage e.g. Dropbox and Googled rive, thus reducing space needed in the device to hold these files.

Skitch: This app helps communicate in real time with other workers. Visual communication is made possible and files can be edited on Skitch and sent as pictures, text etc. to anyone.

The hassle of explaining editing and changes in business plans or ideas is made simple and there is no time wasted in meeting with people.

Square: This app for iOS facilitates workers with iPhones and iPads to take credit card payments on their iPhone. It is a very low cost and efficient method of accepting payments from customers on a mobile device—not to mention secure. These are just a few of the top apps available for smartphone users all over the world which can be adapted for business use with ease. In the end, it all comes down to meeting your company's needs and satisfying the requirements of your employees and customers. So whether you use any or all of the above-mentioned apps or others that are equally capable, the key is making it work for you.

The market for applications is vast, and as the rule of the market dictates, where there is demand, someone is bound to come up with appropriate supply. This is why in today's day and age, the market is full of companies that specialize in designing software that help facilitate a particular company. Thus, any individual specific needs of a firm that the marketed apps do not fulfill can be catered for by such specialized institutions that will do so.

The Best Smartphone for Business Needs

Of all the mobile devices on the market today, the cell phone undoubtedly wins the prize for enabling the highest degree of mobility. . Cell phones more portable than tablets and laptops, Cell phones are used to store large amounts of business data and are used by individuals and businesses all over the world. Smartphones these days are extremely intelligent devices that help in simplifying and enhancing simple business tasks for the users. Smartphones help raise worker productivity through use of apps and make tasks less tiresome: enabling them to take on more work than they could have otherwise.

The question, then, is which smartphone is best for meeting the needs of your business? Unfortunately, there is no easy answer to this question. Because technology is ever-changing, it would be impossible and even foolish to try to award one phone the status of 'must-have'.

While not completely current, a careful analysis of the top contenders is performed below:

Apple iPhone 5

The iPhone 5 is the latest addition to the series of Apple's iPhones which have been released every year with minor improvements in each model. This phone is the best choice for all Business-related use for several reasons. Even though it falls into one of the highest price categories, it is undoubtedly the best smartphone available for executive use. The physical changes aren't the only improvements in the phone as several other factors make it the top choice for all organizations that do not have a low price cap on their budget for smartphones. Apple has always had the best app selection and seems to facilitate the business class better than its competitors. It provides more apps than any other brand or operating system in the It literally has an app for everything in the world of business.

Samsung Galaxy SIII and Galaxy SIV

Samsung's Galaxy SIII was released in 2012, but still seems to be a good choice for those who prefer to use the android interface Over Apple. The phone has become more reasonable in terms of price and offers all the best features and apps that a High-end Android should. The upgraded version Galaxy SIV which was released in 2013, is larger in size and has a high definition screen, but basically the same operations as that of an SIII, which dominates in regards to being used as an executive phone. The price gap between the two phones is huge, but the SIV provides business users with Samsung KNOX, which is used to provide security to business data. So if companies do not have a tight budget, they may seek to upgrade from the SIII to SIV.

The apps on Samsung Android phones are good for business use, but fail to top the variety and precision that most apps limited to Apple have to offer.

BlackBerry Q10

The BlackBerry Q10 is one of the only two devices that have been released with BlackBerry 10 operating system. It has a fully interactive touch screen and a full QWERTY keyboard, which makes for ease of operation in both areas. The defining point of BlackBerry Q10 for business organizations is that it comes with BES to protect the company's official data and records. The phone may be good for mobility and constant communication with staff, but that's about it. The phone fails to give the user a good app selection, which as previously established, is extremely important nowadays to boost worker productivity.

Samsung Galaxy Note II

The Galaxy Note II is a great release by Samsung: In spite of the fact that it has been on the market for a while, its sales are still considerably high. The phone is larger than the Galaxy SIV but smaller than a tablet. This phone is meant to serve the purpose of a tablet and a smartphone all in one. The large screen and easy viewing of data is a plus point and users find the Android interface easy to use and helpful. It is preferred by most over Apple's iPad due to the fact that it has the capability to serve as a phone and a tablet. Tablet and a phone all in one.

How to Choose the Best Smartphone for the Workplace

Smartphones are considered essential in the workplace and every firm needs to have a clear-cut list of things they want in a smartphone for their employees. Since a business is incurring the cost of providing the smartphone to its workers, it should ensure it gets maximum productivity from its use in the workplace (ROI should be optimum).

The factors that should be considered when deciding on the smartphone that best meets the needs of the workforce are as follows:

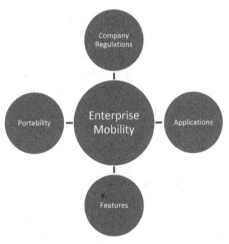

1. Portability

The most essential requirement for smartphones is that they should be portable and easy to carry from one location to another. A phone should not be too fragile like an HTC phone, but also not too solid like a Nokia (their slow working is compensated by the hard and almost unbreakable exterior). Moreover, the phone should be able to operate basic internet commands, which are essential to the mobility and availability of the user to the business.

2. Company Regulations

For the company's IT department to not face a sudden jolt from a completely unknown device being deployed to all workers, the company must keep in mind the devices that are compatible with the current IT setup of the company. If a smartphone is completely alien to the company's IT department a complete rearrangement or re-training of the IT department will have to be done, meaning higher costs for the business.

3. Applications

Considered a great helping hand in raising worker productivity smartphone applications have been necessary for employees. Think about it...if thousands of dollars are being spent on a cell phone, it should provide apps that facilitate the workers in the most optimum way. Apple phones may be the best platform to choose for a firm that wants to raise worker performance as well as satisfaction and the reason for this is that the best apps are available on iOS and the wide variety is also available only on this platform. Apple has apps to facilitate workers in almost every field, for every task from the basic to the most complex. Hence, Apple's iOS are the most popular decision even over the less-expensive Android due to its quality of apps and variety. The thought process being, why choose either quality or quantity when you can opt for both?

4. Features

Once the above-mentioned needs are met, the firm may also consider the specific features of a smartphone. Given that a number of smartphones provide A business with what they desire, the one with the best screen resolution, camera quality, and sound features is likely to be the one chosen

Finally, keeping in mind the factors to be considered when selecting phones for your mobility enterprise system, Apple's iPhone 5 is the ultimate choice to go for. Its standard operations and timely software updates ensure no lags in work and the flawless operating system with the widest range of apps ensures efficient and productive working regardless of the user's location in the world.

Top Ten Benefits of Going Wireless at the Workplace

The need to move fast with an ever-growing business world is crucial for all companies to be able to compete and survive while remaining profitable. That is why ensuring you have the fastest, most reliable Internet connection and speed is a necessity. The transition from traditional connections to DSL and now Wi-Fi has been a profitable one for virtually everyone. Having wireless access to the Internet at the workplace poses no threats to the company as long as a tight security policy is maintained. Wi-Fi uses a wired network to transmit Internet signals over wireless frequencies. Computers and devices in the workplace or in the network can use the wireless connection to communicate within the organization, collect information from the Internet etc., provided that they are in the workplace's Wi-Fi range.

A strong IT department ensures the maximum advantage to be taken of this facility by protecting important business information and data. Among the numerous benefits of adopting wireless technology in the workplace, the following are the most dominant and most common among business organizations in different sectors of activity.

Enterprise Mobility

Every firm craves enterprise mobility and for good reason. The need for mobility is Essential and a wireless network is greatly helpful in facilitating this requirement. The wireless network can help enable mobile and tablet users to say integrated with the company when they are in the office or on the move. Essentially, when a worker is using his tablet at work with wireless network availability, he will have the information he needs with him in his tablet when he leaves the building or geographical 'location'. Wireless networks greatly facilitate enterprise mobility devices.

Better Communication within the Organization

The network can be set up in various locations of the organization. For example, if a 'single' network were set up, the costs of enabling access in every corner of the workplace would and the ROI would be slow. With wireless networks, a simple router and booster can help provide and enhance wireless signals all over the workplace, hence enabling access to the network or internet in far off places in the company Moreover, voice calls and video calls from inside or outside of the workplace can be taken from tablets and mobile phones This is not possible with a wired network. Isn't it plain to see that a wireless network is the only real way to internet efficacy?

Safety Benefits

With wireless networks, innovations in technology have greatly reduced the risk of theft or unauthorized access to business networks. Setting up a VPN or encrypting data for the company would help increase safety benefits in terms of reduced risk of theft or intrusion of important data.

Decreased Need for Hardware

New and updated versions of printers and scanners have an option to connect to computers and mobile devices wirelessly. The major benefit from this, aside from the monetary savings it brings, is the fact that several employees can be assigned to a single printer or scanner through the wireless network rather than needing multiple workstations.

Cost Reduction

We just mentioned one cost-saving factor of a wireless network, but there are others, as well. The cost of materials and supplies for a wired vs. a wireless network are considerably more than the cost of access points, routers and boosters for a wireless network.

Better relations with customers

The company can ensure real time responding and troubleshooting problems of consumers through chatting and conference calls by using any communication device powered by the wireless network. This will enhance relationships with the customers and make them feel more valued.

Little Added Cost for New Employees

When a new employee is hired he can simply be added to the wireless network, which is of minimal cost compared to setting up a separate wired access to the company network or internet for him/her.

Real Time Checks on Inventories and Stock Lists:

A wireless network is invaluable for connecting the different departments/divisions of a business. An employee working in manufacturing production can keep a check on stock lists and inventories of raw materials and finished goods while at the factory or production unit. This is just one example of how wireless connections interconnect every segment of the production process.

Secure Communication

Using a wireless network, all the devices in the workplace can be secured through a VPN for the company and secure communication of sensitive business data can be carried out in the organization.

Easier to Identify Technical Faults

In a wireless network, there are few areas where the hubs, switches and ports are located. In the event of a network breakdown or problem in connectivity, it is easy to identify the cause and the location of the problem, as the areas to search for it are not as numerous as they would be for a wired network. The network is unlikely to suffer from prolonged breakdowns and technical faults and any problems can be addressed in a minimal amount of time.

Cloud Based Mobile Applications

For enterprise mobility, devices that make use of applications is common. As mentioned before, applications have a great role to play in raising worker productivity. Using smartphone and tablets is common nowadays, but no device has unlimited data storage. How then, is a worker expected to perform at his optimum when on the move when the device he is provided with does not have sufficient storage for the company's vast stores of data and details? Yes, the employee may have access to the internet, but how can he access company data and keep it aside for his official use? This sorting and saving of data is not possible for large Businesses with

Large data banks that need to be mined through ERP software for the use of each employee.

The answer to this is what is also called Next Generation Computing, or Cloud Computing. Cloud computing basically means setting up an account on the internet space, which acts as a store for any kind of data the user wants to store and keep safe for future use. The need to purchase expensive an extendable hard drive is reduced, and through keeping this online store for data, the individual ensures that he can access this data from any source as long as he remembers his login details for the cloud service.

Upon reading about cloud computing, one will wonder how this would help an individual who wants to access company data from a mobile device. This is simple. The companies that created cloud computing software have quickly and intelligently adapted to changing business needs and have created their own apps for smartphones. In more official terms, mobile Cloud computing is done when the information is not stored in the phone's memory and all the processing of the information is done in a space external to the phone, which is called the cloud.

These apps are said to work best on Android and iOS platforms, and save employees a great deal of work and hassle when it comes to procurement and storage of official data. With other applications available to open, create or edit existing official documents, cloud based applications are essential to many aspects of work.

A few of the best cloud apps available for smartphones in the market include the following:

- Evernote
- Spotify
- Dropbox
- Google Drive
- Cloud App

These cloud-based applications are the most common in the corporate world, and facilitate large amounts of data storage and mining greatly in organizations without the need for expensive portable hard drives and storage devices.

Most Useful Tablets in the Workplace

Originally created to be partners for computer use, tablets are now completely replacing computers and laptops in the business world. The reason for this is the improvements in technology. Simple tablets that started out as subsidiaries for computers

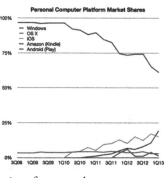

now possess nearly all characteristics of a normal computer.

Tablets have taken over most of the market share for PCs, as can be seen in a graph that represents a market share for Windows PC and different tablet platforms, which are available in the market.

It is no doubt that Apple's iPad is the most dominant in the market. This is because of Apples friendly interface, higher quality performance and the loyalty that Apple customers have towards the company and its products. It is true that many companies even choose to invest in iPads over smartphones for their employees, as it ensures that the device is used mostly for business and in a secure environment and responsible manner. But is it really the best tablet available?

A tablet by Toshiba called Thrive is said to be better than the iPad in most aspects. It does not associate itself with high costs of accessories, and presents the user with technical such as ports for HDMI, and USB devices. This is something not available in which requires costly extensions to be purchased for the use of such devices. It is feared that the iPad is still a subsidiary of a general Apple computer and cannot substitute it. However, its competitors seem to have no problem in doing just that.

What must not be forgotten is the fact that Apple products have the best apps in regards to worker productivity. This 'ultimately' Apple's iPad the greatest edge over all other tablets available in the market.

Another device giving the iPad a run for its money is the Galaxy Note. This is a "phablet" comprising both the SMART PHONE and a tablet. This and the fact that the Android Play store offers many apps Apple's, or close enough to make no difference in reliability and functionality, is making it a very close second… and coming up on Apple to take over the lead.

After having briefly discussed the best tablets available in the market, it is useful to discuss why tablets are so successful in the corporate world. Tablets are Lightweight and hence very portable (adding to enterprise mobility), provide good battery life and easily connect to wireless networks and perform integrated tasks upon the user's demand. Workers can access their email, corporate information and data through their tablets and use it to make their work pattern less tiresome. Moreover, collaboration is enabled as workers can create and even edit official documents and print those using wireless connections on their tablets. This is made possible mainly through the use of the ever increasing number of apps available for download and purchase on the platform under use.

With technological advancements and increasing need for mobility, it is true that laptops are being replaced by tablets altogether in some sectors of business activity. The flexibility and portability enables workers to feel less loaded and more comfortable with their choice of tablet that can be personalized to their convenience.

However, even though many different companies are developing better tablets every year and continually trying to win the never-ending race of producing the best device in the world, it must be noted that tablets can never outrun traditional laptops or desktop computers in the aggregate business world. This is due to several factors that are essential to the core business processes.

It is true that an employee in sales or marketing can use a tablet to create sales portfolios, analyze trends, sketch new marketing plans and deliver top of the line presentations and market proposals using their tablets. But can an employee in finance use an iPad to run Enterprise Resource Planning software and generate complex long term reports for the company? It is highly unlikely, even with SAP integrated apps coming into the market continuously. Certain aspects of business activity require high-density information to be mined, processed and results to be constructed in ways, which use software that is not available in tablets.

These tasks can solely be completed using computers with large processors to enable the worker to carry out his task without overloading the device. There will always be certain areas of business where traditional computers will be needed and no innovations in enterprise mobility can replace them.

Another area where tablets fail to provide an exact alternative to traditional computers is executive planning and forecasting. A company's expansion strategy cannot be formulated using data gathered on the Safari browser in an iPad. For core business decisions to be made proper research must be carried out and reports prepared. These activities are too complex to be carried out on tablets, and if tried, would do nothing but reduce worker productivity and increase dissatisfaction levels with the device.

Tablets can no doubt help increase enterprise mobility, but cannot be seen to replace traditional computers and devices in the near or distant future.

The Features of an Ideal App

It has been repeatedly mentioned before that smartphone apps are essential in raising worker productivity and enhance the return on investment for a company in devices for enterprise mobility. With the wide variety to choose from and the many alternative apps in the same category, how to choose the best app is a daunting task. It is good to know about the characteristics that define an ideal app in order to make the right app decision.

Instant feedback to the user – An app should be able to provide a user with real time responses. Apps that have long time lags in providing results to users are less popular among other similar apps that can give results and responses relatively quickly. For example, an employee wishing to generate a graph of sales trends cannot wait for hours for the Microsoft Excel App to finish its work. The result is the user moving to a faster, more reliable app.

Error elimination – Much like autocorrect in most phones, an app should have the ability to inform the user when it notices an error in input. This helps save the user's time and helps accurate usage of the app.

Efficiency – Most businesses or employees who pay for an app do so with the expectation that the app is of high quality. This means the user will have high expectations for the use of the application. This means there is no room for errors in the programming of the app itself. A calculator app giving different answers to the same problems would not be appreciated or used by anyone. Similarly, a GPS app with internet connectivity issues will not be popular among people wishing to use it.

Relevance to Task – An app should be relevant to the task it claims to perform. An app designed to help manage travel expenditure should be up to date with all the deals and promotions available at the destination. An ideal travel app will be one which provides the optimum use of a budget to travel to and from a certain location.

Cost – If two apps of the same quality and purpose are available to customers, they are likely to purchase the cheaper one, as the return they will get out of its use will be the same as that they would receive from the more expensive one. Low cost, therefore, makes for an ideal app because people consider their primary investment to be the tablet or phone on which the app will be installed.

User Interface – The app must be user-friendly the simpler it is to use, the more popular it will be among the users. After all, this is why apps are designed; to simplify tasks and facilitate the users, regardless of what use they put the app to.

Sales in Apps – Markets are not only competitive for fashion and clothing industries. Apple sales figures have shown that when apps are put on special promotions and sales, the number of apps sold actually increases greatly in the market.

To keep in mind such factors about an app would not only help users select the best application available to them, but would also facilitate developers to produce apps that can help users meet their requirements more efficiently. Thus, both should be taken into consideration when buying and selling an app.

Chapter 5

Enterprise Mobility Strategy

Develop Your Mobility Strategy

The movement towards greater enterprise mobility is motivating businesses all over the world to push their boundaries and look for all kinds of mobility solutions in the hope of providing their customers with the best possible services and giving their employees a productivity boost. Mobility is truly blooming as a concept in the last three to four years and businesses are now able to pick and choose when it comes to choosing a solution that works best for them. Mobility solutions are also within reach, which is why the switch to mobility has become swift, 'and is' Businesses to upgrade themselves to the next level in no time. Despite all the advancements, 'however,' every solution at hand is suitable for your business and diving headfirst without a Pre-meditated mobile strategy can leave you dissatisfied with the results.

One reason for wanting to go mobile might be that your competitors are doing so. Yet, this is merely the tip of the mobility iceberg. There is a lot in store for your business in the long term. Moreover, there is the added benefit of value addition to your enterprise in both measurable and more subtle, immeasurable terms. A mobility solution founded on a well-designed mobility strategy can bridge the chasm between what is anticipated and what is eventually attained.

With the right start, you can go a long way. So start by brainstorming over the following questions.

- What are your enterprise's mobility needs? What do you expect to achieve as a result of enhanced mobility? How do your expectations help bring your enterprise's goals into line?
- Do you want to boost revenues, retain loyal customers, and increase your brand-awareness or a combination of all three?
- What do you wish to achieve in the long term?

What group or groups will form your intended audience? Do these segments within your audience have varying needs? If so, how will you design a solution to cater 'to' group's unique demands?

- How will you leverage existing assets such as information gathered from social media and your current mobile site? Can you possibly use this information to avoid duplication of work?
- What limitations do your current budget and infrastructure pose?
- Do you really know what your employees' needs and abilities are? Perhaps you should gather more information about what they expect from a mobility solution, what they needs want the mobility solution to address, how adept they are at using technology and what suggestions they have in mind.
- What are the expectations of personnel at different tiers of the organization? How can their needs be answered with mobility?

- What learning aids do you need to utilize to train your staff? This decision will need you to take the eventual users demographics, reach and location into consideration.
- How do the returns from developing and implementing a mobility solution compare with its overall costs?
- How can you utilize statistics and data to present a convincing case for going mobile? How do other alternatives fare in comparison?

Once you have chosen a mobile learning vendor, it is imperative that you schedule an appointment with their enterprise mobility expert to discuss how their proposed solution can be tailored to the demands of your organization whether more customizations are possible and what the constraints are. It is also advisable that you contact your potential vendor's previous clients to learn first-hand what working with this vendor did for their business. Ask for feedback on multiple facets of the solution such as the solution's user-friendliness, design, functioning, security features as well as any technical glitches that surfaced pre, during or post implementation.

An all-inclusive mobile strategy must give due consideration to the different modes of connecting with customers using smart phones and tablets. This includes text messaging, SMS/MMS, mobile optimized websites, marketing campaigns, coupons, geo-location services and applications. Making use of all possible modalities will ensure that you cover all audiences and their different needs.

Here is some food for thought for budding mobile entrepreneurs regarding the different possibilities at hand.

Create a Mobile Optimized Website

You might have typed in a website address and found yourself redirected to a mobile optimized sub-domain of the company's website. For instance, you might have typed in 'abc.com' only to find yourself re-routed to 'm.abc.com'. What happens is that once you type in a website address, the site detects that you are trying to access it via a mobile device and hence re-directs you to its mobile optimized version. These versions are often simplified to the basics without delving into extraneous details so that the website loads quickly and the consumer gets exactly what he might need on the fly.

Or a mobile app...

From games to talking dictionaries and baking recipes, apparently there is a mobile application for anything one can imagine. To make a suitable application, you need to know your audience inside out. Do they spend long periods of time on your website at a stretch or would they want access to more precise, streamlined information instead? If your audience falls in the latter category, chances are that an app will suit your business's needs best. Apps that audiences find useful such as What Sapp or App Cache Cleaner, or fun (Angry Bird and Candy Crush enthusiasts know how addictive these can be!) do roaring business.

Recently, Amazon offered discounts to consumers who used their app to scan barcodes to compare prices.

Offer a premium version of your App

Begin with a free version of your app with the possibility of letting your users access a premium, paid version with bonus features. This monetization strategy has been utilized by games (again, recall Angry Birds and Candy Crush Saga) as well as instant messaging applications such as What Sapp, Tango and Skype, amongst others. For example, you can access limited emoticons on Tango in the basic version but are required to pay subscription charges to use the others. Similarly, What Sapp is a free application for the first year of use and requires a small subscription fee thereafter.

The underlying assumption here is that once your users have begun using the application, they will develop such a strong affinity for it that they will be willing to spend money to avail special features or to continue use. The tactic also allows businesses to discriminate between different market segments.

Mobile Coupons, anyone?

Coupons can often encourage consumers to buy items they would not insert 'buy', but most of today's millennia's would not take the pains of carefully clipping out coupons from magazines or newspapers. To get around this problem, more and more companies deliver mobile coupons straight to their clients' smart-phones. These coupons work just like their paper based predecessors- show them to the cashier who will 'then' the barcode just like an ordinary coupon - giving you access to special offers and discounts.

With GPS and other location sharing applications such as Google Maps and Facebook becoming all the rage, customers can be alerted about sales and special offers when in a certain retailer's vicinity. A bargain is hard to resist when it is so tantalizingly close, don't you think? Customers agree, as Jewries Mobile Audience Insights Report revealed 50 percent respondents voting for location-based coupons and location-specific ads.

Make your campaigns and ads mobile!

Mobile marketing gives businesses the previously unencumbered opportunity of interacting with their customers instead of a monologue. This means that campaigns and ads can be tailored in response to user feedback, providing users with precisely what they want, when they want it.

Mobile advertising pushes enterprises to design crisp ads that hit the spot without beating about the bush. For this reason, surveys reveal that mobile ads outdo internet ads, with the former being five times as effective.

Reflecting on these questions before implementing a mobile campaign is a worthwhile investment since it can help you prepare for possible problems well in advance, allow you to opt for the most cost effective solution available and reap the benefits of your implemented solution at the earliest possible.

The Integrated Mobile Strategy

The terms m-commerce and mobility have become very popular recently. Especially since the introduction of the Apple iPhone in 2007 followed by the tablet PC in 2010, mobility has become the talk of the town. Mobility has been hailed as the new IT and is being regarded as the discriminating factor in the years to come between whether your brand survives today's hypercompetitive markets or withers into oblivion.

There is no doubt that mobile commerce is becoming increasingly popular, especially amongst today's youth and the Gen Y. With most of the American population owning both a smartphone and a tablet, the way people look for bargains and shop has 'been' for good. According to a report published by Gartner, a third of smartphone users browse the web to shop and 3 percent will eventually purchase things online. Already there is evidence to suggest that more and more people will switch from laptops and PCs to handheld devices as the primary means of connecting to the internet.

Wise businesspersons know that they must hop onto the m-commerce bandwagon if they want to keep pace with their competition and excel in its implementation if they wish to surpass it. This requires an investment of both time and effort but even despite their willingness; one can sometimes feel a little puzzled about where to get started. The following four are suggestions on how to formulate your business' integrated mobile strategy:

Why not create a customized app?

A website might not always be best suited for your business and a mobile application might be a better choice. You can employ the services of an app developer, buy an off the shelf app or use one of the many app creating tools available online. These tools differ in their subscription fee and the amount of programming proficiency that is required. Whether you choose a mobile optimized website or a mobile application, what ultimately matters is the quality of user experience it offers. A app or a website that stalls in the midst of a transaction will irritate your customer and will reflect extremely poorly on your company's name. Often it will cause irreparable damage to your brand's image. So the need to maintain your application cannot be stressed enough.

When creating a mobile friendly version of your website, do not forget the screen size!

It may sound like stating the obvious but when browsing the website on a mobile device rather than on a laptop or PC, the customer has access to a far smaller screen. Cramming too much data on a mobile optimized website can make the screen look cluttered and cause inconvenience to the customer.

While a version of your current site is what you should have in mind when developing a version of your website, trying to include each and every piece of information is inadvisable. Remember, the screen size is a constraint when it comes to mobile devices. Therefore, it is best that you provide only the most important information such as contact information, stock status and the best offers available. Users on the fly will typically not want to delve into the company's history, so such details are best left out. You may even want to consider including a link to your primary website in case the user feels that the need for more detailed information.

After the website has been tried out by your clients, they will provide feedback on where they felt the website was lacking. This feedback should be incorporated to improve subsequent versions.

Quality Assurance is critical!

Once a user has a poor experience navigating your app or website, chances are strong that they will abstain from ever using it again. Additionally, their lack of satisfaction can lead to word of mouth negativity which will ultimately impact your reputation.

With so much at stake, it pays to take the time to do this right. Perform extensive surveys, test and debug the final version of mobile websites and applications repeatedly. And put yourself in the place of the end-user. Use the app as though you were a customer or client...not as one who developed the website or app. When you take the time to understand that when you take the time to understand that the average user does not have the same tech savvy or technical expertise, you devise a site/app that can be used by all kinds of customers without frustration or difficulty.

Publicize and Advertise!

There is no point in making a mobile application or website if your customers never see it! If your company is going mobile, let your consumers know. Advertise in glossies and newspapers. In addition, since most of the end users will be Gen Y millennial, the best way to reach out to this segment is to publicize the upcoming mobile optimized website or application on social media websites, such as Facebook and Twitter. Another idea worth considering is sending out messages on your existing customer's phones with the website link mentioned so that customers can directly log on to the site without wasting a minute.

This is by no means an exhaustive list. In fact, it barely begins to scratch the surface of all the wondrous opportunities that lie ahead for your business if you implement an integrated mobile strategy. Investing in m-commerce can boost sales by retaining loyal customers, as well as attracting new ones. M-Commerce is an incredible opportunity to capitalize on because there is rarely a moment when customers separate from their mobile devices. Whether at home or commuting, they will surely have a phone on hand. Making your brand's presence felt on mobile devices is likely to be the best thing you can do for your business in these times.

Maximize the Value of Mobility in Your Operation

The previous sections discuss your first steps towards greater mobility: designing and implementing a mobility strategy. The natural, obvious progression is to maximize the value of mobility in your operation once you have a mobile strategy in place. Here too, the services of a well-reputed mobility consultancy can help you a great deal. The consultants there can guide you in striking the right balance between increased flexibility in your employees' work environment while ensuring tight and impenetrable security. Most companies offer to carry out an in-depth survey of the business to inspect the extent of the impact of mobility and identify possible loopholes. Moreover, mobility consultants can ensure a smooth transition to mobility.

Once a mobility solution is implemented, it is crucial for that business to keep track of its performance. This can be done by developing metrics, tools and benchmarks to aid you in gauging how your mobility solution is faring and taking immediate action to remedy any problems that arise.

To top things off and ensure that you remain ahead of the curve, your business will need to devise a far-sighted go-forward plan. What milestones in mobility will you expect to achieve in the future and how will these pan out? This helps you chart your progress.

A consultancy can also help make your existing technical environment conducive to the implementation of a mobility strategy. For starters, the consultancy will conduct a roundup of your current situation and your planned environment. Next, it will try to resolve system integration concerns, assess how ready your operations are for the impending mobility transition and aid you in taking steps that will ensure that your adopted technologies work in tandem with your organization's business processes and vision.

Finally, you need to equip your personnel with adequate know-how to be able to make the most of your deployed mobility solution. A reputed mobility consultancy can help your employees brace for impact by laying out a program management plan and a roadmap of how to mold the management to make a smooth, hassle-free transition a reality. Since every organization is unique in terms of its need and challenges, most consultants will be willing to tailor a solution to address your requirements instead of doling out a generic, one-size-fits-all solution.

If you get it right, a mobility solution can help you get the maximum value from existing investments by ensuring that your business software reaches mobile users as well. The flexibility of working from any location at any time of their convenience can dramatically increase your employees' productivity. Also, your customers are bound to be all smiles when they can access you with greater ease and expect a speedy resolution to their queries. Finally, mobility has certainly come a long way in recent years as far as data security is concerned so entrepreneurs can bid adieu to sleepless nights spent worrying about their data ending up in the wrong hands.

Case Study

Samsung Smart School fuses education with mobility to form a mobile educational solution that enables students and teachers alike to take advantage of a wide variety of educational tools on their smart phones and tablet PCs. Access to the latest school information, vital pedagogical tools, attendance rosters and class participation trackers now rests in the hands of every student and teacher. It is an established fact that these tools, when deployed efficiently, make lessons more engaging for students and help them perform better.

With the facilitation of two-way communication between institutions and students, learning can be much more interactive and fun. Therefore, learning can now be an active pursuit instead of a monologue with the teacher delivering a lecture and students robotically noting it down. Increased flexibility and connectedness resulting from mobile solutions can lead to positive effects that percolate throughout the education system.

With a Samsung Smart Solution implemented, teachers can share content from their tablet or PC with their students' mobile devices. The blackboard seems to be on its way out with the

E-board taking its place, which means static 2-D diagrams can be replaced with animations, digital media and 3-D depictions, which can be rotated in space. By employing these sophisticated tools, teachers gain greater control over their classroom-in fact with natural world phenomena coming alive in the classroom, you can expect students to be spellbound. It is a no-brainer that when children are having fun while learning, they absorb more of what is being taught to them.

The dynamic Question and Answer feature provides students and teachers with instant feedback allowing both to spot weak areas and possible room for improvement. Teachers benefit immensely from the class attendance and participation tracker since the time spent tabulating data manually can be spent in teaching instead.

Students often fall behind when they are unable to access their teachers to discuss problems. This gap in understanding can accumulate exponentially, leading children fall back further and further and eventually losing interest in lessons altogether. Thankfully, the Samsung Smart School comes with a real-time Question and Answer feature. This is such a simple and easy solution that it is a surprise that nobody thought of this earlier! When posed with a problem, all a student needs to do is post his problem his tablet. The problem will become visible on the screens of the teacher and peers tablets as well 'as on the' E-board. The teacher can then take notice of the problem and resolve the query either orally or in writing. Missed the teacher's response the first time around? No need to panic, the Samsung Smart School Solution records the teacher's response and allows all students to access it later, whenever need be.

The benefits of switching to the Samsung Smart School are many: promoting an interesting, stimulating and interactive work environment to strengthen concentration, encouraging participation from even the most introvert pupils in class, and instant support from peers and teachers. Moreover, students no longer need to carry those bulky schoolbags from one place to the other, every textbook they can possibly need can be compressed into their tablets' memory.

How to make your mobility efforts sustainable

The past ten years have seen a marked change in how we define mobility. A few years ago, mobility pertained to very discrete, field specific applications. However, with the explosion of handheld communication devices alongside the less obvious explosion in the availability of 3G connectivity at very low prices, the world has witnessed a new, unprecedented wave of innovation taking place. What is important to note is that this innovation is not always immediately visible to the IT department of a business. This is referred to as shadow IT, where people do things with their own devices to speed up work and improve performance that the core IT department might be unaware of or in some cases, might even frown upon. A classic example of that is the popular Drop Box-its massive use caught most IT departments by surprise.

The fact of the matter is that if we were talking about creating a mobility strategy six or seven years ago, the discussion would be limited to a discrete set of problems but if we were to talk about it now, we find ourselves in a completely different environment and therefore very different problems. This is why the need for a completely different methodology arises.

After years of working, mobility consultants have arrived at what is now called the Sustainable Mobility Strategy. This strategy is rolled out in three distinct phases. To begin with, there is the crucial kick-start phase. As the name suggests, this is the phase where the implementation of a new mobility strategy kicks off. This activity might sound onerous by the sound of it but does not necessarily have to take much time. Nonetheless, it does need to get done. The kick-start phase comprises two stages. The first being the grounded research or discovery phase. This is where the consultants talk about what the organizations goals and motivations for mobility are. Bear in mind, however, that restricting yourself to just the managers in an organization can cause you to miss a great deal of detail about what goes on so aim to get comments and ideas from personnel at all tiers. If you get down to the right questions and are able to strike the right discussions with the individuals having hands-on experience in running the business, you will be rewarded in two ways.

The first thing you will get is a good description of the problems they are facing and what they want solved. In other words, access to the situations in which your employees feel mobility can help. Another important thing you will find as a result of this interaction is how people in the organization are already working-how are they getting around the system. This is vitally important because while the IT department might say that users within the organization must not use, say Drop Box or Google Maps, for any number of sound reasons, people will continue to use them. Your organization can benefit from that if it views this as a benchmark rather than a challenge. Knowing what mobility practices are currently in use can help you understand what the employees within the organization are technically prepared for and also what they can be expected to demand from the IT department in the near future. The grounded research stages quite literally gives you access to hundreds of real life use cases.

Now in the past, IT would have gleaned one or two of those cases to make a case study and then deploy it. That is still seen in a few business organizations but is essentially a highly flawed approach to mobility. When organizations do that, they end up wasting both time and money, but more importantly they end up with a plethora of very disconnected solutions that will simply not meet the business's needs. Over time, the business will turn into a tattered mess.

This calls for the second stage in the kick-start phase, known as the generalization phase. At this stage, consultants will gather all these distinct user cases, bring them together and categorize them broadly. In most organizations, this boils down to three or four main user cases. This exercise allows the organization to generalize all of the many problems into a few main problems. This consolidation allows you to get a bird's eye view of what issues are involved and allows you to solve each problem in turn.

The phase that follows is the structural phase where the actual rollout of the solution takes place. Since it would be naïve to say the process of mobilization ever actually ends, this structuralization phase is an ongoing process. The challenge here is that many organizations try to rush from understanding their organization's architecture in terms of generalizations to strategy. That is another recipe for disaster. The right way to circumnavigate this is to form a governance process so that strategy can be framed in the light of actual needs and the decision-making processes of the organization.

While you may assert that your organization does have a governance system in place, consultants often find that mobility projects tend to be smaller than your traditional large government frameworks but larger than the agile projects that you have. They sit somewhere in the middle. More importantly, since you are dealing with innovation or the concept of shadow IT (whether you like it or not!), you need a governance framework that is very as well as agile and fast enough to make business decisions without getting bogged down. It must also not be so lightweight that you end up in a tangle later on. Therefore, governance must absolutely precede strategy.

The structural phase involves identifying the critical issues within the generalizations and then deploying upon those. Finally, you get to implementation, which is an ongoing activity. Implementation becomes much easier once you have done all that groundwork. In fact, organizations that use this methodology will often need only one or two products to solve a significant number of their use cases. A good case in point is electronic forms. While a number of firms still use manual form entry in the field, many generalized forms products can be prototyped and developed within months or in some cases within weeks. These products can actually solve a vast amount of the organization's structural problems.

Now the deep interconnection and how it is going to fit within the architecture has also been taken into account because you have not only faced the issues of generalization and governance, but you know how everything will fit together within the rest of your system. How this will all fit together with the rest of the system. Hence, the concept of "sustainable mobile strategy." This kind of strategy is a very appropriate, pragmatic and relatively quick process to deploy and get up and running.

4 Tried and Tested Change Management Methods: Making an easier transition to increased mobility

Ever since mobile technology became ubiquitous, the impact of this enhanced connectivity for businesses has been undeniable. IT managers or CIOs want to use this increased mobility to give their organizations a productivity boost while increasing user satisfaction, all at the lowest costs possible. This transition to mobility, however, is a little intimidating at first glance but it need not be. Adopting the following best practices can help make the mobility solution deployment process more efficient ensuring a hassle free rollout?

Try to identify the needs and problems of your mobile employees

The deployment of a new mobile strategy will often be driven by the organization's IT department's motivations and targets. Often this agenda consists of objectives such as making IT processes more efficient or making the best use of data resources.

However, such isolated thinking that is detached from the organization's workers and their needs can be a recipe for disaster. It is therefore crucial that CIOs and mobility consultants make it a priority to understand the workforce's requirements and woes before a strategy is implemented.

Ideally, IT personnel should accompany field workers in their real work environment and conduct focus groups so that they can feel the pulse of the employees' problems and difficulties. By doing so, you can see how where the organization can avoid re-inventing the wheel and detect inefficiencies alongside getting critical, un-intercepted feedback from the personnel who will eventually be using the mobility solution.

An added benefit is that you gain the trust of your workforce by taking them into confidence right at the start.

Begin with a trial version of your final mobility solution

After the initial kick-start phase where you get information about the organization's needs and possible solution paths, start by implementing pilot programs instead of one full-fledged final solution. The feedback from your pilot program is priceless. The 'treated group' can experience working with the mobility solution and convey their feedback along with comments and suggestions to the IT department. This small-scale implementation can give you the opportunity of keeping track of information such as how frequently repairs or service was needed and the impact on productivity.

This information can be used to present a strong case for greater mobility to the higher tiers of management and finance in the organization.

Choose a vendor that offers post-implementation support

When choosing a mobility vendor, stick to one that offers support and services after deployment. All new things take time to adjust to and it should not be surprising if your employees take time to be acclimatized to the new devices and related software. If your vendor provides on-site assistance and support once you have bought their solution, the time required by your organization's IT department for training and helping users with the transition will be greatly reduced.

Provide training for all end users

Once your mobility solution vendor has reviewed your organization and come up with a suitable solution, devise a systematic implementation plan that includes employee training as a vital element. This is essential if you want end users to maximize the value of the purchased solution and utilize all features.

To begin with, consultants advise announcing a particular day to inaugurate the plan for the entire staff with training. To encourage use and gather support from even the most skeptic individuals on the team, have individuals who participated in the pilot program step forward and share their experience with the new users. Chances are that good words about the technology coming from their peers and colleagues will carry more weight in the eyes of the staff. Next, Compliment these user comments with stats and figures on how the pilot program led to improvements in the efficiency of the select groups.

Time invested in discussing the mobility solution with your workforce is time well spent, as it will foster acceptance of the technology and speed up your return on investment.

These tried and tested methods will help your whole workforce, including both digital natives (Gen Y millennia's who opened their eyes in a digital word) and digital immigrants (those who grew up in the pre digital world make the best of the deployed mobility solution. If you stick to this plan, you are bound to save on mobility costs, garner organization-wide support for participation in the mobility technology, and improve employees' personal experience while being considerate about their privacy and constraints-while preserving the security of your data and keeping support costs to a minimum in addition to creating value.

Mobility: Its Impact and Integration into the existing Operating Model

It might be hard to believe but not too Long ago, the IT department of an organization reigned supreme when it came to controlling the technological aspects of the business. Select employees were allotted desktop PCs that were completely managed by the IT department with no interference or assistance from anyone. And if you were bestowed with a Blackberry, you were among the hierarchy of the organization and expected to treat the device with the respect and reverence it deserved. And like the PCs, the Blackberry was fully serviced and supported by the IT staff.

Then along comes 2007—the year the first iPhone made its debut. Suddenly, the smart phone did not seem nearly as formidable and out of reach as it once was and more and more people wanted and obtained one of these incredible devices. The response of

IT departments in the business world was neither positive nor friendly, but despite their reluctance, they had to accept that mobile computing and the IT department itself had changed for all times to come.

With the advent of mobile devices that obviously weren't going away, the process of technology assimilation into a business completely reversed. The IT department, which, up to now, had the luxury of researching the various kinds of tools at its disposal, testing before implementation, and installing these tools prudently considering what best fit the security tools already in place, no longer had this option. Almost overnight, the newfound ubiquity of IT meant that users had already chosen devices based on their personal tastes and requirements and almost by accident, discovered that this had the potential to increase their productivity at work. Suddenly, IT departments had this swarm of new devices cropping up all over their organizations and being used to access email, cloud based business applications and even confidential company data.

That is where the situation becomes tricky. Not to doubt the employee's intention, but personnel can often be naïve regarding the workings of mobile devices introducing massive loopholes in the corporate security. That means that unchecked usage needs to be curbed and the kind of devices and applications used, as well as how they are used, needs to be governed by a policy framework. Bring Your Own Device policies, programs and products can help cut hardware costs and facilitate a flexible work environment while mitigating the security and privacy risks of employees using their own devices at work. Despite its crucial nature however, only Thirty-nine percent of organizations that have BYOD in effect are utilizing Mobile Device Management Systems, which shows how they are trivializing the matter.

That translates into Sixty-one percent or almost two thirds of all companies that might potentially be jeopardizing the corporate security by practicing negligence towards MDM systems.

Virtualization is a practice borrowed from the networking that is now being used as a solution in this case as well. It essentially involves mobile devices to be classified as different cases and each is managed and secured individually.

An alternative to BYOD that IT departments prefer as a way to get around security risks and management issues is COPE-Capitalize each word. This practice enables using virtualization and as the name suggests, the device is owned by the company but used by the employee. Since it is easier to regulate and standardize, COPE is seen as more manageable than BYOD but employees often complain that it is comparatively inflexible.

Considering the breakneck speed with which new technologies are being introduced, IT departments find themselves walking a tightrope. While all businesses want the benefits of higher productivity, customer connectedness and employee satisfaction, they must also juggle these with the obstacles of integrating new apps with legacy systems and installing new devices. Moreover, despite the rhetoric, mobile security solutions are not always chosen based on what works best in the interests of the organization and its employees. The choice is heavily influenced by vested interests of the top tiers of management.

The big question is what can IT professionals and business managers do to avoid missing the mobility train? To maintain a competitive advantage, businesses must update their knowledge of the latest mobility trends at all times. It's a bitter pill to swallow but the fact of the matter is that what is considered cutting edge technology today will be discarded as obsolete and clunky tomorrow. Staying ahead will require you to keep yourself informed of the latest devices, the new threats and the most sophisticated solutions without fail. However if you keep at it, you will become proficient at telling MDM apart from MAM and terms like BYOD, BYOA and BYOX will start rolling off your tongue like a pro.

To take an analogy, enterprise mobility is to business computing what the invention of the telephone was to communication: revolutionary. If your business has not embarked on this thrilling journey already, the time to do so and take advantage of this change is now.

Managing Organizational Change

The fact that enterprise mobility has revolutionized business practices right to the core is undeniable. Ever since the iPhone hit the market in 2007, we have experienced fundamental changes in lifestyle and work style, from how we obtain and utilize information to how we interact amongst ourselves and with our favorite retailers. Since new technologies are hitting market shelves at breakneck speed, businesses that want to remain ahead of the curve must keep on their toes and adapt quickly if they want to capitalize on the gains from enhanced mobility.

As technology became cheaper, business owners found more and more employees opting for smart phones and tablets and using these devices to increase their productivity. Like it or not, most organizations realized there was no getting around the phenomenon and hence the Bring Your Own Device initiative came into practice. Since most mobility initiatives result as an aftermath of already popular employee practices or demands, proper governance structures to support mobility are either delayed or rather flimsy. Moreover, little thought is put into bringing mobility initiatives in line with the business's overall goals. This unplanned, spur of the moment approach to mobility might work temporarily, but prolonged negligence towards developing a sustainable strategy will lead to a divergence from the business's goals and little or no increase in value.

To make the most of mobility practices, businesses must take the time to evaluate in what ways these practices are predicted to affect the existing operating model. How will current business processes evolve in response to greater mobility? Organizations that indulge in this assessment exercise and invest the time to weave mobility into their operating model are able to extract the most out of their mobility initiative especially in the terms of greater customer satisfaction, shorter transaction time cycle and a considerable improvement in employee output.

The shift towards making your enterprise more mobile can be tricky to manage but established management best practices can make the transition smoother for all stakeholders involved. The idea is to break up the task along three dimensions and address each dimension in turn. The three basic areas that will be affected most as a result of your mobility effort are the people in your organization, the processes and of course, the technology used. Managing changes along each of these fronts separately can make the task a lot less daunting than trying to tackle it all at once. The ensuing sections describe how this can be achieved.

Change Dimension #1: People

When your organization moves towards greater mobility, those affected can be classified into two categories. The first category is that of internal people, which includes your employees, and contractors that have been provided a mobile device by the company or those who are bringing their own devices to work and using business apps provided to them. The other category is that of external people; people not employed by the organization but who interact with interact with the business on a regular basis. These include customers, potential customers, and suppliers.

Internal people, especially workers whose main capital is their knowledge, such as architects, engineers and, can expect a dramatic change in how they work, their schedules and where they perform their tasks. To prepare for this shift in their workforce's work style, organizations must become flexible towards accepting various kinds of work patterns as well as managing employees working remotely.

External people such as customers now take easy and constant access as a given and expect that from the businesses they connect and collaborate with. Organizations must keep themselves abreast of the latest mobility trends to meet their clients' expectations.

Change Dimension #2: People

The biggest impact that mobility has had on businesses is in terms of shrinking time delays and location latency. Businesses are beginning to realize that even with basic business apps put into common practice, they can transcend the constraints of time and location when communicating with customers and business partners. The gains from deploying enterprise mobility technologies can only be reaped to the full if the business process underlying it is prepared to tackle it. There are several case studies and examples illustrating how businesses are experiencing high productivity because of investing in the right mobility tools. Retail salespersons can check inventory levels and process transactions with handheld point-of-sales devices and retailers such as Amazon have recently introduced apps that reward consumers with discounts for using them to compare product prices anywhere. There is even news of candy vending machines "informing" stock- refueling trucks passing by when their Inventory is running low! What is important to observe here is that your business need not be inherently digital in nature to make use of mobility tools- all it takes is a keen eye and a little creativity to make your current business processes efficient using enterprise mobility.

Change Dimension #3: Technology

As technology becomes increasingly ubiquitous, businesses need to become comfortable with a wide range of mobile devices, apps, form factors and location-based data quite literally taking over. Integrating these devices and apps into the existing enterprise structure requires the management and the IT department of the organization to put their heads together and devise a mobile device strategy. Three dimensions will need special focus: security management, device management and application management.

While continuing with password protection, data encryption and antivirus software as you did with the desktop PC, mobile devices will require added security measures such as remote lock and wipe, data fading to erase critical information that has not been used by the network for a pre-specified time as well as in the flight data encryption. Device Management pertains to the physical control of the device or a policy regulating what course of action needs to be adopted in case of device loss or theft. Application Management has to do with ensuring that instead of all workers using their own version of the app each requiring different management and update tools each connected to a one multiple servers, you plan a mobility migration with a holistic strategy put into effect so that multiple devices and apps can be managed from a single console with a single set of management tools.

Moreover, forms and reports relying on Create, Read, Update, and Delete CRUD interface are quickly becoming obsolete so application design will have to take a more task-oriented approach.

If managed well, mobility has the potential to unlock a world of possibilities for the organization. However, poor management can lead to problems spinning out of control to the extent that it becomes a management and support nightmare. As the maxim says, well begun is half done? Undoubtedly, having a streamlined, well thought out integrated mobile strategy can help you get the highest ROI on your mobility migration.

Case Study: How Cisco managed change

An enterprise with 300 locations in 90 countries, Cisco has become synonymous with the concept of IT. In 2008, Cisco's IT Network and Data Center Services (NDCS) shifted from a silo legacy model with little information sharing across departments and a lot of duplication to a six-stage life cycle methodology. This lead to an optimization of its IT services and resources with a simultaneous reduction in cost and a measurable improvement in productivity across five distinct metrics at all levels of the organization.

As we described in the last subsection, good management involves simplifying the task at hand into smaller, manageable parts. Cisco's Lifecycle methodology governing its organizational structure involves six phases: Prepare, Plan, Design, Implement, Operate and Optimize. These stages form a continuous cycle designed to optimize Cisco's functioning.

Prepare

This stage involves envisioning the anticipated change taking all factors, requirements and associated technologies into consideration. It also involves justifying the need for change implementation by building a strong case for the financial rationale behind the decision. With a roadmap in place, businesses can keep costs low and anticipate potential problems in advance.

Plan

Assess the organization's current state of affairs regarding the network, security and adaptability to assimilate the new mobile technologies overall. Evaluate the company's finances to ascertain how the project's costs will be borne from start to finish. Lastly, the IT department needs to formulate a roadmap identifying available resources, possible problems, assignment of responsibilities and the strategy needed to meet the deadline within budgetary constraints.

Design

Elaborately describe the Insert-- making sure it aligns with business goals and the organization's technical requirements. Moreover, new custom applications need to be designed in a way that they can be integrated with the organization's existing infrastructure.

Implement

This is the phase where new devices and applications are integrated while maintaining network availability and performance. The major stages involved are installation of hardware and applications, configuration, integrating them with the legacy system, testing the newly installed technologies for bugs and getting all systems operational. Once the system is up and running, IT staff can be trained to increase productivity and reduce latencies.

Operate

The aim of this stage is to run the system in the most manner without compromising on performance. The IT department uses several metrics to monitor the working of the network to improve service quality, minimize glitches or other technical disruptions, work around outages and uphold high standards of security, availability and reliability.

Optimize

Continuous improvement is forming the foundation of the Lifecycle methodology so the managers are always on the lookout for methods to improve the existing process to gain a competitive advantage. As the business, its requirements, its goals and the market evolves, the network must evolve in step so a good business cycle is naturally perpetual.

After deployment in the customer services sector, this Lifecycle methodology to change management was put to practice in IT and proved highly successful. How successful? The results speak for themselves. In the words of Cisco NDCS Vice President John Manville, "By moving from a traditional technology, silo-based organizational structure to a Lifecycle-based model, we were able to improve our operational metrics considerably. Our number of cases decreased by approximately sixty percent, and our time-to-repair to get clients back up and running has decreased by almost seventy percent". That is outstanding progress by any standards.

Increase Customer Connectedness with Enterprise Mobility

Field service representatives are an integral part of an organization's customer service team. In addition to providing technical support to the customers, a field service representative allows for the communication of information from customers to the firm and vice versa. It therefore comes as little surprise that enterprise mobility solutions can be deployed to enhance the productivity of field service organizations- thanks to smart phones and tablets that allow instant sharing of information and markedly shrink operational costs.

In today's hypercompetitive market environment, having a unique product is, on its own, insufficient to make your business stand out from the crowd. Neither does it pay to engage in price competition, as the only result of that is to pull profits down. Therefore, more often than not, excellent customer service is what discriminates between successful enterprises - those that barely break even and those that collapse. Timely recognition of the importance of customer connectedness can and does make all the difference when it comes to creating sales.

It is an established fact that enterprises that provide the best customer service experience retain more customers, experience more repeat purchases from existing clients and more new customers because of word of mouth publicity.

The pertinent question for any business therefore is how do you replicate the superior customer service experience of successful organizations? Basically, it involves resolving your consumers' prime concerns and knowing what makes your consumers happy. To keep your customers satisfied, your organization must aim for the following three goals:

- Correctly diagnose what is bothering your customer and resolve the issue at the earliest possible, preferably on the first visit
- Practice punctuality when meeting for appointments. The customer must know that you value his time
- Provide the requested service on time

These aims cannot be attained by a field service representative if he works in isolation, cut off from the business and denied access to critical back office operations, other field and back office personnel, asset managers and most crucially, customer data. With the right mobility solutions and devices, all that data can lie in the palms of your field workers' and managers' hands. Moreover, as mobility becomes increasingly sophisticated, an odd blackberry or two provided to a few managers in the hierarchy but that are now obsolete, can be debilitating for your organization. What you need instead is an overarching Enterprise-Grade mobility strategy that makes efficient and optimal use of all the most up-to-date mobile trends.

Since the term 'Enterprise Grade Mobility' has not appeared in this book before now, a description is in order to clarify things. Simply put, it is the use of mobile devices such as smart phones and tablets for business applications. Enterprise applications differ from consumer applications in several ways such as data manipulation/management and the level of complexity but the core distinguishing feature is the extent of connectivity. Applications geared towards consumers can perform precisely one hob or function and there is no communication between that app and other apps or systems. Business applications, on the other hand, have the distinguishing feature of being able to communicate with other systems and applications so that there can be exchange of critical information on the go. The best mobility solutions are designed to predict the needs of your workers and are able respond to them across a wide variety of situations.

Let's return from that slight detour to the current section's topic... how enterprise mobility can lead to increased customer connectedness. If you can effectively bring together cutting edge technology and qualified personnel, your customer service can benefit immensely. Here's how:

With a mobility solution deployed, your employees will have an all-round: virtual-view, so to speak, of every customer. This allows your employees to provide your customers with optimal service due to the fact that they are able to access information about the customers' product usage and complaint history on the fly.

Your field workers get instant access to all the information they need on their own without the hassle of calling office personnel for every minute detail. Have a query about the availability of items in the inventory? Need the help of a colleague while you are out on the field? Tapping a few tabs on your mobile device screen is all that it takes to get all these (and several other) tasks done.

Businesses need timely feedback from their customers to identify areas that need improvement as well as to continue with the practices that customers are satisfied with. With the aid of a mobile device, the customer can instantly do so using a survey application. And unlike paper-based surveys, where a few staff members need to be dedicated to tabulating and recording feedback, today's savvy applications do all that and more with a single click.

Mobility enables you to streamline the entire process of directly interacting with the customer, known as 'customer-facing' in business terminology. In the past, there was a considerable lag in the collection of data from the customer and its communication to the concerned department. Moreover, keeping track of a complaint and measuring issue resolution progress was cumbersome. Thanks to mobility, however, each step along the way can be traced accurately and in real time. Having easy access to this information instead of rummaging through endless piles of Paper makes customer facing incredibly swift.

Whether it is getting acknowledgement of receipt of samples on the spot, accessing patient history while on a visit to the patient's residence, finding information about the availability of machine parts in the main warehouse... without a doubt, mobility has permeated every facet of how the customer services department conducts its business and has definitely done so for the better. Complaint resolution times have reduced dramatically and feedback and suggestions on products/services can be acted upon quickly.

Greater mobilization of the customer services department can take your business a long way therefore, all businesses whether large or small, will do themselves a huge favor by acting fast and investing in the design and implementation of a holistic mobility strategy.

Increase Employee Productivity with Enterprise Mobility

Logic dictates that to get the highest return on your investment, each employee working in your organization must give his or her maximum output. The question is how to entice and provide the equipment and environment which will allow employees to perform optimally.

There is growing evidence that the ubiquity of mobile devices in the work environment at all levels, starting from the top tiers of management to the workers out in the field, has given businesses a productivity boost of Forty to seventy percent. According to another estimate, employees in the U.S. today are working eight more 'a week' than they were a decade ago. Effectively that amounts to working an extra day during the week!

How has mobility increased employee productivity? Research indicates that the following factors have made the biggest contribution:

Firstly, the policy of Bring Your Own Device (BYOD) which is becoming a staple part of most businesses is a key factor in making employees more productive. Mobile devices have become an integral part of people's lifestyles and allowing employees to use these devices to get tasks done at work at their convenience lets them get more done in less time.

Enhanced mobility results in constant connectivity to the office, which ensures that not a single minute goes to waste. Employees can utilize their travelling and commuting time as well as the time spent waiting for appointments and meetings to get office work done. Whether it is sending an email or preparing a presentation for an upcoming meeting, mobile apps mean that a lot can be done on the go.

Working from a cubicle from nine in the morning to five in the evening will soon become obsolete because today's mobile enterprises provide employees the flexibility to work at a time and location of their choice.

Mobile Devices Provide Easy Access to Data

Online dashboards and video conferencing software allow individuals in distant locations to arrange meetings at on short notice; saving both time and commuting costs.

Need access to your patient's last blood test report? Or the level of stock in your inventory at at any given time? Sophisticated business apps provide real-time statistics about your customers and business activities.

Nothing improves employee performance as much as effective communication and timely feedback from colleagues as well as clients. Field workers today can get instruction and customer feedback at the point of sale allowing them to track and review their performance.

Finally, pooling in ideas is possible even when team members are in disparate locations- even on different continents. Almost nothing impedes the flow of communication since brainstorming ideas via sharing multimedia content is possible wherever you are.

For all these reasons and more, mobility is becoming firmly entrenched in today's businesses especially with the practice of BYOD gaining acceptance with most employers. The mobility trend exploded in 2012 and most business gurus see progress along the same lines well into 2014.

SECTION 3 –

The Scope of Enterprise Mobility

Chapter 6

Enterprise Mobility Management

Identity Management

Imagine a typical day at work. Chances are that at the main entrance of your office building, you will hold your smart card under a smart card reader to let yourself into the office premises. Once at your workstation, you will log into your computer with a distinct password. Depending on what country you live in, you might also have a health smart card in your wallet that stores all your personal data, which was once entered on paper based insurance forms. Then of course, there is the credit card right next to it.

Each of the above examples are associated with the idea of Identity Management one way or the other. Within a large system of many individuals, such as a company or even a whole country, the need to manage the different identities arises. In the context of enterprise mobility and enterprise IT, the concept can be narrowed down to all the processes involved in establishing and administrating the positions and access rights of individual network users. IT managers deploy ID management systems to increase security and productivity, control access to sensitive information and minimize costs, downtime and work repetition.

In the corporate environment, a quality ID Management system ensures that there is one identity per individual. The identity does not remain stagnant throughout the access don't capitalize. As the employee's role within the organization changes, an Identity management system must be designed to provide service authorization to individuals with a genuine, officially sanctioned business need and remove those authorizations when access is no longer needed. That is why the digital identity of an individual needs perpetual maintenance, modification and monitoring during an employee's tenure in an organization.

Identity management systems allow administrators to do all that and more. They can be used to modify a user's extent of access, follow their activities and to implement policies when they are enforced. An ID management system ensures adherence to the company's and the government's laws and regulations and proper deployment allows the company to manage its employees' access across all functions and levels of the enterprise.

The company has the choice of selecting from a host of identity management technologies such as password management tools, provisioning software, security policy enforcement applications, apps that allow reporting and monitoring and Identity repositories. Since online identities are becoming increasingly connected, both the incidence and cost of identity theft have risen considerably. This has led to a demand for more sophisticated tools to verify basic credentials- passwords alone are not sufficient anymore. Federal agencies in the U.S. are anticipated to shift towards digital data control, biometrics and cryptographic authentication to tackle the situation.

Within enterprises, software suites (functionally related collections of computer programs) are increasingly being deployed. These can be used throughout the enterprise to manage a number of security modalities from credential administration to smart card and digital certificate management.

"Identity Lifecycle management" has recently become the hot term in ID management. It is an umbrella term that encompasses all the processes and technology used to provision and de-provision user accounts, management of workflow, administration of identities, credential management using passwords, digital certificates and automated smart cards, compliance as well as creation and management of roles within the organization.

Case Study: The deployment of Identity Management in a Higher Education Institute

The Premier Research University located in the USA boasts of a population exceeding 15,000 including students, full time and remote faculty members as well as research collaborators located within the academic fraternity as well as in the corporate sector. To improve service quality to its 75,000 users, the university employed the services of Fischer International, an Identity Management firm, to provide applications and target directories.

Why did the need for ID management arise in the first place? Owing to the ubiquity of computers, each department at the university whether academic, administrative or research oriented has its own separate IT department in addition to the central IT organization. Since the IT departments are disjointed in nature, there is no synchronization of data across departments leading to considerable data duplication. Moreover, gaining access to a department or restricting access to it based on the most recent updates was a long and winding task. Other problems included exorbitant costs of user information management, long lags in the provisioning of services, which affected productivity negatively, and obviously, user dissatisfaction because of inconsistencies in accessing computer resources across departments.

When Premier Research University first discussed its business needs with Fischer International, it laid out the following end goals:

1. Address cost and security breach concerns of the organization by deploying a centralized identity management system. The policies of the new framework must be delegable to all departments.

2. Ensure that each department's directory and application are up to date with the most current user information available so that users can have a more convenient experience with regards to privacy and resource access.

3. Provide each department full control over maintaining data on its users' identities and validating authorizations in synchrony with the central IT department.

These targets were achieved by employing the following solutions:

The Fischer Provisioning software was implemented by the central IT organization to create an enterprise directory that both the central IT department and the other departments could use. This directory lists details on all accounts and their users, including students, alumni, administrative officers, guest users etc. The enterprise directory is also the data source that feeds into all departmental, central and functional applications deployed all over the campus. Smaller departmental IT organizations have the choice of utilizing the central data as it is or to create a separate account by modifying existing data.

Another characteristic of the Fischer Provisioning Application is to identify changes in the directory at the central or departmental level in real time so that master accounts are brought up to date instantly and independently of where the actual account is.

Fischer Provisioning allows vast amounts of data to be stored and processed. To quote figures, 100,000 provisioning transactions are processed each day and the enterprise directory has 1.4 million objects in memory.

What was the business value of this identity management solution?

- Allowed cooperation between departments to increase productivity, reducing costs, mitigating possible risks, enhancing service quality and making processes more transparent without a reduction in the departmental IT organization's control over resources or processes.
- Enhanced workflow flexibility while keeping administrative efforts and costs to a bare minimum. An example of this is the process of withdrawing access to classrooms and labs at the end of the school term with a single click, while access to other areas that remain in use such as dorms and libraries is retained.

As a result of improved data sharing, collaboration between research communities and departments became easier. The university became more conducive to learning because of facilitated access to scientific and technical resources.

Application and Data Lifecycle

Application Lifecycle Management (ALM)

While converging to a single definition of Application Lifecycle Management (ALM) might be difficult considering the diversity of perspectives taken by different people, you will hear the term a lot and will come to realize that it's an important idea worth becoming familiar with. ALM is often taken to be synonymous to Software Development Lifecycle (SDLC). This viewpoint is perhaps overly simplistic since ALM encompasses a lot more than SDLC.

To explain in layman terms, ALM includes the whole time span during which an organization is spending money on the application, right from the time the application's idea is first conceived till the time the application ceases to have any utility for the company and its use is discontinued. To gain a precise and helpful understanding of ALM along its entire breadth, one's view of it must be equally broad and flexible.

It helps to visualize an application's life cycle as that of a living organism, punctuated by certain turning points and milestones along the way. A certain need within the organization pushes the IT team to try to conceive an application that addresses the need efficiently. Once created, the next stage is successful implementation of the application within the organization. This involves consideration of budgetary, technical and time constraints while endeavoring to maximize productivity and efficiency. In the end, as new technology arrives and the app becomes obsolete, it loses business value and is aborted and removed from use within the organization.

For ease of comprehension, one can think of ALM as having three distinct elements: Governance, Development and Operations. Governance is an ongoing process that stretches over the entire lifetime of the application. It includes all the managerial activities that keep the application up and running particularly executing critical decisions and managing projects. The Development phase occurs right at the start following conception. It includes all the steps involved in creating the application and leads up to the time when it is deployed in the organization.

Development occurs within the organization periodically, typically when the application upgrades to the newest version available. Once the application is put to use within the organization, a host of activities go into running and managing it successfully; these activities can be aggregated together and called the operations phase. This phase extends from application deployment till the time the application's use is aborted-the activities in the operations phase are carried out round the clock. Now that the reader has a bird's eye view of each stage, we can delve deeper into what each stage entails.

Organizations take one of two approaches in this regard. The simpler approach involves a member of the technical team itself rising to the role of project manager or someone from the project management unit becoming assigned to work with the development team. In more regimented business environments, the central project management office enforces governance procedures itself. Once the application becomes part of the organization's application portfolio, the Application Portfolio Management framework makes use of several metrics to measure the application's performance and to conduct a cost-benefit analysis. It also ensures that each application has distinct, non-overlapping functions to avoid duplication. The metrics generated by the APM system give critical information on how frequently an application is accessed and how the business benefits from its use so that decisions such as whether use should be continued or stopped, whether the application needs an upgrade or replacement.

Governance is the only aspect of Application Lifecycle Management that remains constant throughout the Application's lifetime. The right app with poor governance can result in large losses, with the business not even recovering the initial investment made.

While software development and ALM are far from synonymous or interchangeable, development of a custom application is an integral part of the ALM. Once the business case for an application obtains approval, the work of the software development begins.

Software development basically involves the same process being looped over several iterations. Each repetition involves bringing the application closer to the optimum by redoing the design and then testing. There is a debate over whether this iterative style of development is better than the traditional methods deployed earlier but regardless of each methodologies pros and cons, the iterative method is gaining popularity amongst app developers. After the software is developed, the Software Development Life Cycle is said to have reached its end. This is followed by the application's deployment in the organization. Deployment does not necessarily entail that the application development process ceases altogether. It is quite the contrary, in fact. Every time the need for an update arises or a new version is required, the Software Development Life Cycle must begin anew. In fact, the development of new versions of software is sometimes even costlier than the original.

Once an application is installed, it needs constant monitoring and management to perform optimally. Operations, too, are closely linked to Development. For instance, as soon as the application development is completed, a plan needs to be devised to deploy it within the organization. Deployment is a crucial part of operations. Post-deployment, the Operations personnel keep an eye on the application constantly for as long as it is in use within the organization for possible improvements and correction of bugs. Each updated version that is deployed involves the operations team as well.

Thus, the three elements of ALM, can be thought of as the legs of a three-legged stool-take away one and the stool loses balance and topples over. While getting each stage perfect is a difficult task, the real challenge lies in getting all three elements work in synchrony. The importance of striking the right balance is perhaps best explained by means of an example. An organization that nails every step of the application development process and operations but fails in establishing good governance will definitely generate poor ROI. On the other hand, the governance part might be dealt with extremely well with rightly identifying business needs and then pinpointing the important stakeholders, but weak operations such as scarcity of resources will lead to disastrous results.

Understanding the fundamental nature of each aspect of ALM can help organizations maximize the value of applications. This can be difficult to tackle since the ALM tools available at present leave a lot to be desired especially in terms integrating all three elements into one connected system. Yet it is irrefutable that getting ALM right boils down to taking all three phases together to form one holistic system.

Data Lifecycle Management (DLM) ·

A concept related to Application Management is that of Data Management. Data Lifecycle Management refers to the policy framework that deals with all the data that an information system contains throughout the course of its life: beginning from the point in time when the data is first created by entry into the system right up to the time it is removed. Data can be classified into various levels according to stipulated criteria for optimal management. This activity can be automatized by making use of several DLM software's available in the market. Data can be retrieved more quickly from newer, more advanced storage media. For this reason, data that is current and repeatedly accessed by users in the enterprise is stored on newer, quick retrieval media, while that which is accessed seldom and is less important in nature is stored on slow retrieval, inexpensive media.

Data Lifecycle Management and Information Lifecycle Management are sometimes taken to be synonymous terms;

however, there is a fine line between the two. DLM is far simpler, dealing with a narrow range of file characteristics such as type, size and age for categorization and storage. ILM products, on the other hand, are far more sophisticated in terms of the kind of search operations that they can undertake. An ILM product can retrieve data that meets a more complex criterion, e.g. a particular customer's contact details rather than a file accessed on such and such date.

The enforcement of data compliance regulations means that data management has become more important than ever. Instead of taking the narrow-minded view of DLM as a single product, data management specialists suggest imagining it to be an overarching policy framework that simultaneously governs the procedures, practices and applications involved in the management of an organization's vast data bank.

Deploying New Mobile Applications

Our World is becoming increasingly as a result the consumerization of IT. More and more businesses are turning to cloud based solutions for business agility, cashing in on the fact that more than half of the United States is using a smart phone. When you add tablet users to this, the figure goes through the roof.

At work, the pervasiveness of smart phones, together with the growing acceptance of BYOD (Bring Your Own Device) policies, have resulted in the enterprise market making a sizeable chunk of all mobile service revenues- Early thirty percent. This figure is expected to undergo dramatic growth in the near future, surpassing the rate of growth of consumer based revenues by two times as much. VDC research predicts the field service applications market to grow at eleven percent per annum.

Moreover, a report by the Aberdeen Group about mobility and its impact on field service, published in 2011, revealed that organizations deploying a mobile app solution for the reconciliation of part requirements for truck stock realized an advantage of ten percent in first time fix compared to organizations that did not.

The variety of applications available in the market, even when you leave out bespoke applications, is confounding enough to leave you dizzy. While each is advertised to be the ultimate in employee efficiency and productivity-you cannot possibly deploy them all or sample each one by one. To make the right choice, discuss these five key questions:

- How versatile is your mobile application solution vendor's application portfolio? Do the apps they offer address your business' needs or will a customized application benefit your organization more, even after considering its additional costs? If a customized application seems to be best suited to your requirements, can the vendor configure available applications to address them?
- How easy are the mobile applications to configure? Do they need extensive programming know how or can the average user, with no programming skills, configure them as well? Will you have the option of combining the functionality of two or more applications if that caters to your organization's unique requirements better?
- Are the applications compatible with a variety of commonly used devices and operating systems? As consumerization of IT gains hold, the demand for device agnostic applications is increasing. Such applications are not bound to any single device and can be used with different mobility devices with ease. In the face of the BYOD trend, having inflexible applications would entail very little collaboration and crushing pressure for application support from the organization's IT

department to each type of device being brought to work.

- Can the proposed mobility solutions be integrated with existing legacy back end systems such as Enterprise Resource Planning (ERP) without hassle? Is the mobility solution vendor willing to extend support and guidance at the integration stage?

- Will extensive upgrades be required every time you want a new feature or function in your mobility solution or is there a way around them? In the event that upgrade is deployed, will previous applications cease to work and require complex upgrades or will they keep working?

All five of the above concerns are readily addressed by enterprise mobile app stores. New functions and features can be obtained by simply downloading new applications without the entire system requiring an overhaul. These app stores also allow users to choose, configure and integrate any number of applications with ease.

Any new application deployed needs to be flexible and adaptable enough to be modified to answer the vast variety of challenges and situations a business finds itself in on a day to day basis. In today's hyper competitive and fast paced business world, no business affords the luxury of waiting for the IT department to write a new program for each business scenario or for a better version of the application to be released. The ideal app needs to be modifiable according to the ever-changing state of affairs by making use of simple configuration features. DIY (Do It Yourself) functionality is the name of the game.

Deploying new applications can lead to improvements in both productivity and efficiency in all areas of the enterprise. Enterprise Mobile App Stores are making this simple, since they support a number of fully integral and easily configurable applications.

How to integrate new apps with an existing system

Adoption of cloud computing is being hailed as the most revolutionary trend in recent times. Business owners, thought leaders and customers all want to hop aboard the mobility train.

Cloud computing promises efficiency, speed, enhanced storage and considerable cost savings making it an incredibly attractive option for businesses wanting to extend mobility. The only 'boogeyman' keeping IT managers awake at night is the possibility of new cloud based apps not integrating with the enterprise solutions already in place within the organization. A complete system overhaul is both disruptive and expensive so being able to integrate new applications with already deployed systems is a prime concern.

Scribe Software recently conducted a survey of 920 business leaders which brought forward some interesting findings. System Integration is indeed a valid concern with around ten percent of the surveyed enterprises showing no evidence of integration across various systems. Only sixteen percent of the enterprises achieved full integration across various systems deployed. Integration between ERP Systems and new customer relationship management applications lagged farthest behind with only thirteen percent of the survey population being able to do so successfully.

Cloud based solutions may be user friendly but even they pose considerable challenges when it comes to integrating them with legacy systems and data sources. Among those businesses surveyed, most reported struggling with the integration of CRM solutions with the new social applications-only one in ten organizations can claim to have integrated the two successfully.

New systems require new system integration methodologies but most organizations still rely on custom code development as a means of doing so. The pace at which new businesses must respond means that cumbersome solutions like developing your own custom code and relying on manual data entry methods are more of a handicap than anything else however these practices continue to persist. According to the survey's findings, forty-eight percent of respondents still use custom coding as a method of system integration and about one in three organizations still use manual data entry processes.

Projections based on these results would entail that most enterprises would only deploy their own IT departments to deal with data integration issues rather than diverting towards employing the services of external systems integrators or IT consultants. However the surprising finding is that nearly half the participating businesses demonstrated the intent to adopt hybrid environment support showing that more and more enterprises are rightly identifying the integration of new mobile apps with existing systems as a top priority.

Since more and more businesses want to shift towards greater enterprise mobility as a means of gaining access to real time market data, customer insight and connectedness, employee satisfaction and productivity gains, there is a pressing need to bridge the chasm between the legacy systems installed and the new generation cloud based mobile applications.

This need is being addressed by the software industry with more prebuilt integration solutions and third party platforms being created recently. More businesses are realizing that employing the services of professionally trained systems integrators and consultants is a worthwhile investment.

Businesses can take heart that while the integration of cloud based mobility applications with legacy systems is challenging; it has become a lot simpler than it was a few years ago. The need of the hour is that new integration technologies and skills become the norm. Adopting these new methods will help businesses extract maximum returns from today's powerful cloud applications.

Platform Fragmentation

Most smart phones today are based on the Android platform. Android is a software stack introduced by Google Inc. A software stack has several utilities working in tandem in a predefined sequence to achieve a certain objective and in the case of Google Android; this includes the operating system, interconnecting middleware and important applications for use on smart phones.

The Android SDK is the software development kit developers use to create applications for the Android platform. The programming language the Android SDK uses is Java and the applications are then run on Dalvin.

With these two concepts explained, the issue of platform fragmentation can be elaborated with much greater ease. There are several ix variants of the Android platform in use at present, This concoction of variants can make things very tricky for IT managers. There is a constant threat that apps written with the Android SDK will not be compatible with the ever-increasing variety of the Android platform that different smart phones deploy. This issue is generally called Android Fragmentation or Platform Fragmentation. This means device agnosticism will be compromised as the number of custom versions of Android rise, applications that run on one device, will not be operable on other devices implying that new programs be written for each individual variable. This obviously complicates matters unnecessarily and drives up costs.

There is a multiplicity of opinions about whether the panic surrounding platform fragmentation is justified or whether the issue is simply being blown out of proportion for no reason.

Google has introduced an Android compatibility program to put down such concerns and enhance interoperability. The extent to which this resolves the issue of mobile specific, inflexible applications is debatable.

Fast-paced technological space

Technology is evolving at a dizzying pace- with most users terming it to be as fast as greased lightning. Trying to cope with this rapidly evolving state of affairs while keeping costs at a minimum is a Legitimate concern for business leaders the world over. A recent report published by the Economist Intelligence Unit revealed that almost half of Europe's business leaders have concerns about failing to keep up with the ever-accelerating rate of technology change and realize that not doing so might result in losing their competitive edge in an aggressive market.

While there is no denying the productivity gains that business innovation enables, it is equally obvious that slacking off even a little in terms of employing the latest mobility solutions compared to one's competitors can have a devastating impact on market share and performance. This concern was cited by a majority of the senior executives interviewed and the finding resonated across major economic hubs in Europe, North America and Asia Pacific. As businesses become increasingly reliant on IT, internal business processes must be designed such that they can move in step with the newest technologies available. This section has some suggestions on how this adaptability can be achieved.

Capable business leaders can maintain their calm when things around them are going at whirlwind speed. Your team will already be a little perplexed at the implementation of new unfamiliar technology, so it is of utmost importance that you stay focused. Working to the best of your abilities as a business leader requires being able to unwind and relax. There are several relaxation techniques that will help—depending upon your preferences and circumstances. Unplugging your Wi-Fi device when you go to sleep might sound rather paradoxical in nature, but the fact of the matter is that being wired 24/7 is robbing us of our sleep. Take a walk, hit the gym or relax with yoga relaxation techniques... whatever it takes to may you feel relaxed and renewed; allowing you to maintain your professionalism and output. The most potent tool against the fear of being in unfamiliar territory is knowledge. Whether it is the fear of not having the know how to operate the new mobility solutions or the Herculean task of memorizing acronyms like MDM and BYOD, gathering information from credible resources and reading up will help you gear up for change. As a leader, you need to be aware of the technologies that are deployed in various areas of your enterprise. Leading effectively will require you to keep abreast of the latest research and developments in mobility as well.

As more and more recent graduates join the workforce, business Leaders will face the added challenge of managing a workforce with diverse values and attitudes towards work. You have a growing number of digital natives- Gen Y individuals who opened their eyes in a digital world working alongside digital immigrants those individuals who grew up in a pre-digital world. While the very mention of this existence of generational differences in the workplace might sound compartmentalized—as if we are putting people in boxes by age- the fact of the matter is that the generation gap in terms of adapting to IT is a reality. Therefore, as a business leader, you will have to have an awareness of the issues of each age group's concerns regarding whatever new solutions you deploy as well as a multidimensional, all-encompassing plan on how you will go forward with resolving them. This will help you retain employees of all ages, increase employee satisfaction, boost morale and help spur innovation.

The power bestowed to a leader comes with its responsibilities and share of challenges. In today's hypercompetitive business environment, the foremost of these is adapting to current realities with an eye on what's on the technology horizon in the coming future. If business leaders can truly master that, they can unleash the real power of enterprise mobility and reap astounding gains from it.

Privacy Issues

The easy access to information brought about by mobility is (in regards to privacy) a double-edged sword. Sure, your employees can gain access to the most current data about the business but so can a number of hackers. Mobile applications deployed in enterprises are allowed to draw on sensitive company information such as contact details, emails and schedules. Without installing software to enforce security, this information can be leaked. To mitigate the risk of losing precious intellectual property, IT security departments must have a viable plan in place both for cases of physical loss of mobile devices as well as interception of the system otherwise. This involves going beyond just monitoring employees' mobile device usage.

With BYOD and BYOA becoming increasingly common in the modern workplace, employees can download applications on their devices that may compromise the organization's privacy and allow leakage of sensitive data. Every time you download an application, you agree to grant the app permission to access personal data without much thought into its consequences or even if this access is necessary for the application to function. Despite the assertion made by Google, Apple, Windows and other mobile platform providers about stringent control over the malware and negligence, it is best that you go the extra mile to secure your network and have a privacy policy in place within your organization.

Mobile application developers want access to customer data because there is an inherent clash of interests. They want to assure customers of the best user experience and privacy protection yet there is the opposing pull to share usage data with ad networks to make money. It is these ad developers that pay for what users view as free apps but the situation may not be as rosy as consumers deem it to be. In exchange for free or cheap apps, more often than not, users unknowingly compromise their privacy and reveal data to commercial entities.

Mobile security is typically handled by the system administrator in an organization. He is responsible for implementing the security policy, network software as well as hardware necessary to protect the network from unauthorized access while ensuring that authorized personnel have uninterrupted access to the network and the resources they require to work efficiently. A multifaceted approach that utilizes hardware in addition to monitoring and security software can work best to enhance overall security.

Case Study: Advocate Health Care puts Patient Privacy first with Air Watch

Mobility is becoming increasingly popular in the healthcare industry since it leads to efficiency and speed for doctors as well as the supporting staff. However, because of the sensitive and extremely personal nature of patient information, there is a need to deploy mobile security solutions simultaneously. Quality Mobile Device Management (MDM) solutions make sure that healthcare organizations conform to the Health Insurance Portability and Accountability Act's tenets regarding patient privacy protection. With BYOD becoming the norm, Advocate Health Care's IT department turned to Air Watch for a mobility solution that is both HIPAA compliant and productivity enhancing.

After a thorough evaluation of Advocate Health Care's business needs and prevalent practices, Air Watch instated full device protection, password enforcement and asset tracking solutions. To manage device users, data on device usage across the organization is published and reviewed on a weekly basis. In the case of device theft or loss, the data on the device can be wiped remotely. The MDM software provided also ensures that internal and government imposed compliance standards are met. Each nurse can view lab reports and patient data on her own handheld device instead of waiting in queue to access the desktop at the nurses' station. This does wonders for their productivity. While visiting patients at home for follow-ups, nurses can review critical patient records from the palm of their hands; there is no need to lug around bulky files anymore. Information sharing across departments is swift without the fear of security breaches.

Thanks to Air Watch's security solution, Advocate can provide first class patient care on the go without ever compromising on its stringent patient privacy protection standards.

CHAPTER 7

Enterprise Mobility Solutions

Devising a successful Enterprise Management Strategy in the wake of massive recurrent technological challenges is an arduous task for any organization. Managers, the board of directors and owners have to consider a plethora of factors before deciding what solution to implement. The benefits and necessity of having Enterprise Mobility are doubtlessly essential, but its execution is the real test of business grit.

Enterprise Mobility Solution will engender the approach, the tactics, the methods, components and timings of incorporating instruments of mobility in your system. A holistic methodology requires careful planning and intelligent deliberation. For all the aforementioned challenges, the solutions would carefully consider original business objectives, the objectives of a proposed strategy and find ways to resolve issues in a most cost effective manner, while at the same time keep an eye out for alternatives to a proposed strategy.

Key Features of Enterprise Mobility Solution

- Multiple Platforms - Foremost, an Enterprise Mobility Solution needs to have 4-5 platforms like BlackBerry, Android, iPhone, Windows etc. for market adoption. It must be operable on these platforms so that it could tap into a diverse base of users.

- Build or Buy - Next, the business has to decide whether these platforms will be engineered or if they should be built from scratch. The solution should be and beneficial in the long term for the organization.

- Compatibility in a Fragmented Market - The device should also be functional in a diverse marketplace, and that the business app is compatible with systems that are predicted to come on the market.

- Focused on User Experience - An Enterprise Mobility Solution has to carefully gauge and cater to the needs of the user; such that the app is interactive and attention gripping.

- Addressed Security Concerns - Enterprise Mobility Solution should tailor itself to fulfill the data security desideratum. The business data is sensitive and there should be no inlets for hackers or viruses to put that in danger.

- No Loopholes - In addition, the app should adequately function even where this is no Internet connectivity or limited bandwidth, to avoid halts in the communication process. The features within the app should be well connected so that the user experience is choppy.

- Synchronization - The app should be synchronized with the enterprise back-end, i.e. it should be regularly updated to reduce excessive battery consumption and bandwidth on the device itself.

- The decision between Mobile & Cloud Integration - This decision is specific to an organization, if the invention of an app is cumbersome, a Cloud computing system will pool resources, and provide benefits like increased elasticity and agility for smooth functioning.

- Adoptive - The Enterprise Mobility Solution should be adopted by users; it is imperative for it to be user friendly. It should provide the familiar look and feel of native applications.

- Integration - The Mobility Solution has to be integrated; it is not a point solution to the business concerns, rather it works in conjunction with the IT infrastructure; expertise in both, mobile client platforms and enterprise back-end is mandatory for this.

- Support system - The Enterprise Mobility App has to anchor against a support system of the IT infrastructure and web developers.

- Application Analytics - A successful Enterprise Mobility Solution should be able to track user behavior and observe trends; such analysis should then feed into a business' evaluation of its Return on Investment in a mobility solution.

- Testable - The flexibility, endurance and bottlenecks in the functioning of a mobility solution should be scrutinized by varying bandwidth load, altering the security configurations and testing the back-end infrastructure.

- Future-proof - The solution your business devices should be progressive; it should be easily updated to meet the technological challenges that your organization should have forecast in the strategy-making phase.

Is your Mobility Solution Enterprise Ready?

Please provide the reference of the book in footnote, there are four ways of determining whether your mobility solution is enterprise ready.

1. Understand the end-to-end business process your application supports: Regardless of who purchases your

application, understand how your application measures the criticality of your solution to your customer. The more entrenched the application is in the business process and in the collection and storage of data, the less likely a customer will be to shift from your solution to another. It is also imperative to link the application to corresponding or neighboring apps so that the user does not have to act as a mediating, connectivity layer. If a business fails to do that the users, mainly employees will have to assimilate the different apps themselves that is a costly, time consuming and redundant situation.

2. Find areas to modify your application to increase customer value: It is also vital to add value to the mobility solution with additional product functionality. Provision of related features within the existing app can help achieve this. With the advent of open APIs businesses can expand their operations. Making the discovery of and development to their API convenient, simple and accessible will maximize the chances of garnering new clients and retaining existing ones.

3. Make your application Plug and Play: To enable third parties to use your application, you need to make your API as simple as possible. The best way to accelerate adoption of your application and reduce barriers in time is to enable easy plug and playability of your application.

4. Simplify the customer buying process and user experience: Having a seamless user experience is another important facet. Staying connected with the user from the moment they log onto an application, to the time they complete the purchase, will dispel any feeling of alienation the customer may have form interacting through a device. Availability of simple, relevant information on the screen is also a key to good user experience.

An enterprise-ready app needs verification to ensure that only the employees that are allowed to can access data and go through with transactions. In addition, the important applications should be functional even in the face of no or less network coverage.

The key consideration, whilst determining whether a mobility solution is enterprise ready or not, includes introducing distributed system architecture with redundant components. The deployment should be gradual and inclusive, divided into manageable phases so that users stay on the same page as the organization. Using a content delivery network, a flexible network bandwidth and distributing loads where needed, are also important facets of a trenchant mobility solution. Moreover, tests gauging the performance of the system and device are necessary.

Furthermore, the mobility test standard includes the following benchmarks:

1. Cloud solution - Supports on premise solution or managed mobility models
2. Performs data Partitioning - Keeps enterprise and personal information separate that aid in tracking costs of operation
3. Provides integration to middleware applications
4. Deploys functional apps that the user can readily link to
5. Reports analyzing user trends are prepared to give business valuable feedback
6. Manages expenses by giving early warnings for excessive cost and usage
7. Scalability load testing
8. Affords secure transmission of data within the devices (SAP Mobile)

Mobility is an important component of an organization because it provides flexibility in connecting across departments. It pools information and resources to provide the users with easy access to documents and memos circulating within a business. They can edit work, provide their feedback, and stay in touch with the business round the clock. As a system, mobility cannot be segregated from its foundation, i.e. the IT infrastructure.

Businesses need to evaluate their capacities to ascertain whether the mobility solution devised through strategic planning is suitable for the enterprise or not. A mobility solution will be used daily for a long time; the success of the enterprise is couched in the logical functioning of the app itself. There has to be a thorough appraisal of the potential messaging techniques, the myriad 'of' to connect, the constituents for integration, security and lastly, the steps needed to propagate the use of mobile devices within the organization.

Existing mobile users can give valuable insight into the effectiveness of mobile devices; they can be asked to provide feedback about the pros and cons of the application or the software, its interactivity, user friendliness and problems. The processes that are a part of this mobility solution, for e.g. sending messages, entering data related to customers, keeping records of inventories need to holistically engage the IT development, the mobile computing and the end user. Areas that still require some form of paperwork or documentation should also be defined clearly.

- **Mobile middleware** should allow seamless and continuous access to services throughout the organization. There is a need to provide for device management and accessibility so that once a system is in place it functions automatically, integrating all loose ends in a cohesive manner.
- **Controlled Deployment** is essential for the stage in which an organization distributes and launches its mobility solution. The deployment should not

be hasty or forced, it should occur in stages so that the workforce is not suddenly overwhelmed with mechanical contraptions. The standards, which the users are required to adhere to, should also be outlined. Overall measures should be taken so that the device is well received and effectively adopted.

- **Effective IT support** is also mandatory for every mobile IT initiative. The IT infrastructure in place is the backbone of the mobility solution; it should focus on placing structures that aid the deployment of mobility solutions. The onus of securing the data, analyzing the input from the users and ironing out glitches in the system is placed on the 'shoulders of' department

Create a Great Consumer Experience with Increased Enterprise Mobility

Enterprise mobility reaps multiple advantages for all stakeholders. The B2C (Business to Customer) approach highlighted earlier focuses on invigorating consumer experience. This approach focuses on revenue generation by involving customers in the usage of a device or an app (Roberts). Mobility offers finesse to the customer buying experience and supplements the existing sales by attracting new customers. The following areas are affected by this:

- Sales – Interactive and appealing apps that match the customer's buying ideology attract them. If the app itself if charged for, then the customer will buy it if they can relate to it; if the app is free, it gives customers an inlet into the business itself and with the adding convenience of mobile shopping, chances are they might buy more of the firm's products.
- Expanding customer base – customers multiply based on contacts, for e.g. if the app is an incredible success they

will provide free advertisement by recommending it to their peers. In addition allowing for an easy download of the app through your website is another way customers can carry your business in their mobile. Placing the app in an application store is another way to target potential customers.

- Maintaining a customer base – customer service is a sensitive area where prompt response of a business is important. Mobile app is an aid to interaction; the businesses that uphold consumer sovereignty and give an unparalleled buying experience have a more loyal customer base.

Businesses need to be dynamic and flexible according to their customer base. Knowing your target audience and its profile also provides substantial leverage to developing a strategy that bolsters consumer experience. Consumers are not homogenous; therefore, there is no one stratagem that will fit all business needs. Broadly speaking, a business aims to create a company environment focused on understanding and meeting customer expectations. A good customer experience transpires when expectations are in line with a business' endeavors to create satisfaction. A strategy should be delineated from the customer's perspective. The first step is to define the experience and then to align the business' culture, internal processes, technology and products to create that experience.

Knowing the Customer is a prerequisite, which further requires meticulous data collection that taps into the behavior of the customer.

- **Focus on sales**: Businesses need to learn how, when and where customers buy their products. Consumers are already well informed about different products; interaction with a business is the tipping point of choosing whether to buy or not, thus businesses need to focus on the customers' need for information to support an informed buying decision.

- **Influence:** Within a customer base, there are some buyers who Rule the roost when it comes to influence. They are, so to speak, the popular kids in school. They spend more on products and their activity makes them influencers. Influencers should be identified and careful marketing strategies should be directed towards them. A business app should have analytic instruments that give feedback on buying patterns which a business can use to increase sales.

- **Mobile experience:** With the increased need to have everything in a portable form, customers want to access the business' products and services from different locations. Therefore, providing authentic data about products from different access points is essential.

- **Identify your customer:** The business-customer nexus is complicated and has various points of interaction, for e.g. the advertisement, the customers' appraisal of your products, your sales service, the delivery of products and after-sales service. Businesses need to know exactly what sort of information content and communication the customer needs at each of these points. This is possible if there has been extensive research on the customer base; the multiple identities of customers online have to be tied together to give meaningful forecasts. Regarding these customer communities can be of great help. Allowing reviews, ratings, complaints to be displayed on the website is a covert way to pinpoint customer identities.

- **Measure:** Make sure you create a consistent experience quality across all interactions. At every stage of deployment, the user reaction needs to be assessed and recorded. In the testing phase, an app designed for employees should capture the issues they face or the features they like best. For customers the number of users downloading or buying the app, the trends of their purchase, how many purchases reach a conclusion, which ones are left in the middle; such aspects need to be carefully quantified and translated into app assessment reports.

- **Technology:** Technology should be a link between different areas of the enterprise as well as a component of business machinery. Businesses, employees and the customer should be collinear. The flow of communication should be circular, i.e. businesses need a strategy for customer experience, they need to collect data through social media about customer behavior and then feed that information to the strategists to devise better policies.

Enterprise mobility has altered the buying experience in many ways. Increasingly, businesses are aiming to deploy mobility at every level from browsing of catalogues, to ordering, delivery and after-sale service. Automation using a mobile device reaps benefits in terms of improved accuracy, no human errors, and simplified standard buying procedures that not only augment productivity but also satisfies more customers.

The benchmark for developing an app is the customer preference and expectation. The app has to be engaging, interactive and interesting for the customer to continue using it. Businesses cannot use conventional measures as a yardstick to satisfy well-informed and spirited consumers who seek a combination of work and entertainment. Innovative techniques need to be employed to maximize consumer satisfaction. For example, the mobile app on LinkedIn is more interesting than its website.

For many businesses, modifications in mobile devices have become a source of revenue. Square is a good example of such a firm. Businesses and customers were allowed to make and accept credit card payments using a device in their phones. It provided an additional incentive of low cost and heightened convenience, thereby dissolving the barricade that constrained traditional credit card use. More people can now extend their business and offer services by accepting credit card payments.

Similarly, Near Field Communication (NFC) allows two devices to exchange data if they are within a specified range. It is being increasingly used in public transportation systems, where consumers can use their NFC devices to make transactions. This is perfect for systems where smart cards are already used.

It is important to collate information from certain locations and assess if more services can be provided on those interaction points, e.g. point-of-sale services have been revamped using mobile technology. Target provides its customers with the option of collecting coupons on mobile; at the time of the purchase, they can retrieve and use these coupons using their mobile. Mobile devices allow users to perform small and menial transactions, i.e. check a bank balance, finding the store location and so on.

Through a mobile enterprise system, businesses can connect with customers more regularly and dynamically. The three components that allow ease of transaction and ensure that the IT substructure is innovative, pliant to changes and integrative, are social media, analytics and cloud mobility.

Technology should always be welcome and intelligently employed in business operations. It opens up new avenues of selling products and informing customers. Social media aids in strengthening the image of the business in the minds of the customer through advertisements, schemes, online competitions and offers. In addition, businesses no longer have to operate in niches, they can target a wide range of customers and develop and discover new segments.

Developing an App for Your Business

Designing an app that meets the business' Enterprise Mobility needs is gaining ground. An app will be tailor-made to suit exact needs and it can connect with employees and customers exactly the way a business envisions. Mobile apps contain a personalized vision of the business.

The first question any business should ask itself is, does it need an app? Most likely a business app is an inevitable step, but the size and needs of a business should directly influence the decision of whether an app would even be necessary or simply complicate matters further. All businesses must consider the following.

Target Audience

Knowing about the potential users and whether they have smartphones or not is the first step. Obviously, mobility solutions will differ if they are only connecting the executive tier or if they are for the employees or the customers. Each app will be designed differently. Next, if the strategy is of Bring Your Own Device, then the emphasis has to be placed on the most user-friendly platform or a combination of platforms for building the app. If the business is to provide the device then the cost effectiveness of such a decision has to be assessed.

Budget

Developing a mobile app is expensive, thus a business carefully needs to conduct a cost-benefit analysis and maybe even survey its employees to find out what will they prefer to make the internal system mobile efficacious. The costs for storage, security, web developing also need to be accounted for.

App Content

Mobile apps need to be updated constantly, in order to attract more users, while also retaining the old ones. This stage is crucial to the success of the mobility strategy. With careful deliberation the exact requirements must be penned down together with the prospective goals one aims to achieve. Research is essential at this stage; the business team must work in collaboration with the web developers, thus there has to be a crystal clear vision of what will appeal to the needs of the business. The general vibe of the app, what is displayed on the screen, where the data is stored, are questions answered at this stage. The ability to edit, save and share information among employees is also determined.

Cross-Platform Formatting

This refers to the ease with which one app can be used on multiple platforms, so that the users are not restricted. This feature gives the app a degree of universality and diversity. Users feel at ease that no matter what kind of phone they have the app will be functional on it. However, such a decision is obviously expensive and complicated.

Cost Vs. Profit

Businesses need to evaluate if the net profit would be able to exceed the costs by a reasonable margin from the deployment of the app. In most cases, the initial cost will be more, but gradually benefits like ease of ordering, more sales, fewer errors will accumulate into sizeable profits.

Testing

The app needs to be carefully tested and made secure for usage. Unforeseen issues can be ironed out at this stage. Julia Rogers elucidates the five things needed to develop an app:

1. Clear vision with a stepwise approach

The business objective dictates every decision it makes. A clear delineation of goals is very important to get prompt results. Dave Caste Lenovo believes, because all entrepreneurs love the creative process, an influx of ideas can divert focus from the core concept".[3]

It is also important to leave sufficient room for improvement and updating the app with new developments. Apps can be developed in stages with more features added on to the existing structure later on. Dividing the task into simpler components and concentrating on the core areas, e.g. having a foolproof ordering system for a retailer is a wise approach.

Robert Boyle, digital media strategist, explained that although mobile apps help build and consolidate the business image, create brand awareness and boost sales, the act of developing the app can go hopelessly astray if the goals are not clear from the beginning. Thinking from the user perspective is important. The organization and the user are two parties often times geographically distant; the mobile app has to be a bridge between the two, not an impasse.

2. Let consumer preferences direct the design

---

[3] Julia Rogers. "Building Apps For Your Business: 5 Things You Need To Know." 17 Oct. 2011. Huffington Post. 14 Sept. 2013. http://www.huffingtonpost.com/2011/10/14/building-apps-for-your-business_n_1011681.html

Research is a vital part of any strategy; the business mobility solution can be succinct and simple but it has to be operable. What customers want is what the solution should deliver. The consumer demographic is very important; their age, gender, occupation 'and' area of residence will give insight into the sort of app that will be suitable. Again if there are cost constraints, intelligent designers will make apps that suit almost all age groups, 'but' will only be possible if the app is simple. Estimating what your competitors are doing is also essential. The pristine segments of a market can be identified and the firm will learn from the strengths and weaknesses of its competitors.

3. Make sure your brand name is noticeable

The design of the app should stand out; there are multiple apps in the market. What gives one app an edge over others if its kind? Is it a unique design or feature? Does the name, icon and color scheme stand out from the rest?

4. See if there are existing solutions available in the market

The adage that governs businesses is low cost, high returns. In order to achieve the former, careful analysis of the costs of inputs is required, namely the hardware that will be used, the software, the cost of web developer, legal costs like patents. Next come the deployment cost; testing and distributing the mobility solution, updating redundant features and so on. Building a business app from scratch is not easy or cheap with, but it will have the advantage of meeting the exact requirement of the firm. The facets of business culture e.g. more/less supervision, punctuality, autonomy will be reflected in the type of app it creates. Nevertheless, for small businesses it is not necessary to create an app because of the large overheads involved and the lack of economies of scale.

Yet a mobility solution is a fruitful investment. Hiring a developer is the next stage--one that requires homework on the part of the business a developer would need clear and detailed instructions about what is required. Documentation of elements and features is the cornerstone of this. List similar apps that closely match business needs to give a better idea.

Nigel Hawthorn, director of EMEA marketing at the enterprise mobility management software stated that the renowned pharmaceutical company, Novartis has internally developed more than fifty, along with commercially developed enterprise applications (Burton). These apps cater to different sets of people in various countries. Even the departments within the company are developing apps to meet their requirements. He said, "The key to writing apps that people will want to use, is to be disciplined over functionality in order to make the app as easy to use as possible - not to let committees of marketers run wild trying to cram in more than a limited screen estate can handle".[4]

Case Studies: Innoppl

To offer a mobile shopping experience, a retail store asked Innoppl to formulate a B2C mobile solution for their e-commerce store. The features of this mobility solution entailed the provision of a "shopping cart, product search, product catalog, product description, payment and shipping integration (Innoppl)".

Additional features allow customers to create a 'wish list' of products they want to purchase in the future, they were able

---

[4]   Graeme Burton. "Enterprise Mobility Summit 2013: Corporate Mobile App Development Booming." 11 June 2013. Computing. 14 Sept. 2013.
http://www.computing.co.uk/ctg/news/2274200/entprise-mobility-summit-2013-corporate-mobile-app-development-booming

to locate stores in their vicinity and could contact retailers if and when necessary. The customers could easily make support calls and mail the retailers if need be Change to read: Innoppl also assisted with the integration of the app with a store's back-end system to make it compatible across different platforms and screens.

Another mobility solution was developed for Auto Claim Technology to replace paperwork by providing its workers with iPads to record work in the field. For real time operations, a web back-end was provided for the app to interact with. The choice of the platform, i.e. iPad, was important because workers could take pictures of the automobiles for documenting the model and make repairs. A web portal was also developed to allow management to monitor data and generate invoices and various reports.

The automation process started ex nihilo; workers were trained to use the app and were given a PIN for accessing data. Not only did this app benefit the employees, but it also reduced the time taken to generate an invoice for the client from four days to only one.

For another company, Nixon Eglin, an iPad app was created for salespersons with the features of "customer-lists, address book, register events and geo location displayed on the screen (Innoppl)". The sales personnel were able to submit reports soon after closing a business deal. Previously the customers' records were made manually, which was a time consuming process, but now the field workers could add customers on the go.

Beautiful App and its Advantages

An application is a bridge between the business image and its employees or its customers. Businesses that care about their image and want to innovate effectively will invest in an app that is appealing to the user. The three main varieties of apps are built for:

- Consumer sales for the purpose of entertainment and personal life-management
- Marketing, branding, or customer service (Maxwell)
- Facilitating internal operations, mail and other employee needs

The last two categories consist of apps that are widely distributed free of cost to reach the widest range of users. For instance, Starbucks has developed an app for its customers that allows them to pay through their phone, gives information about discounts and offers, gives directions to nearest stores, allows customers to explore menus and nutritional information. The appearance of the app is engaging and vibrant; options like Drink Builder, where the customer can design his/her coffee. It also allows customers to suggest changes to the app itself and lists job opportunities for prospective employees (Starbucks). Thus, Starbucks personally connected with its clientele.

Similarly, the Target app for iPhone allows customers to plan trips, use mobile coupons to save money, check prices. Its Beta Shopping List allows customers to connect to the Wi-Fi in the store and to add, search or delete items as they go.

Adoption of the App

A major task for a business developing its app is to make sure employees and customers alike find it convenient and enjoyable to use.

For instance, NetApp, a company trying to simplify the process of app adoption and its employees released a "Cafeteria Ordering" app that is different from the usual enterprise apps. The app allowed workers to order their food and subsequently use the app! Getting employees to adopt an app is vital. Apps need to offer entertainment value to its users. Such an approach gathers attention for the app within a business.

Decide upon a strategy for the App

In the field of retail businesses, businesses need to decide whether to use "enrichment" strategies that emphasize on garnering new customers and driving them to make a transaction or an "engagement" strategy that enables customers to increase their involvement with the brand.

J. Crew launched its famous app, Very Personal Stylist, which allows customer to experience twelve style stories and select apparel. The interface is attractive and allows customers to make an appointment with the stylists. Similarly, Gucci's app contains "a mobile point-of-sale program that lets employees register sales, email receipts to customers, access the Gucci Style app and use a translator and currency converter on the spot (Carr)"; Helping to increase sales and brand awareness among sales associates while simultaneously allowing them to offer better service.

According to a study done by research firm Latitude, Seventy-nine percent of shoppers are interested in the possibility of having digital content such as product recommendations, demo videos and virtual "try on" simulations delivered to their mobile phones while shopping in a store (Latitude).

In order to converse with your customer, the mobile app needs to connect with their needs at a personal level. Signature excels at this by offering an app with both an employee and a consumer interface where customers can interact directly with their personal sales associate (Skidmore).

Design the App

Sixty-one percent of people have a better opinion about brands that offer a Positive mobile experience. For this reason alone, the mobile app should be trendy, appealing and unique- something that stays in the mind of the user. Customers clearly state that, "a lot of shopping apps have too big of a footprint; that is, they're too clunky, intrusive, or ad-heavy Compared to how much value they deliver. They need to offer an experience that justifies taking up storage space on [any] phone".[5] The IKEA NOW app offers customers the facility of using a 3D catalogue, which makes their shopping experience versatile.

According to the study by Latitude, the ideal shopping app should:

Sell products and give information- Along with selling products, companies need to focus on offering intangible services like mobile coupons, ways to compare prices and features of products, a system to track down the shipment and delivery of the products and reviews from existing customers as a guide for those who want to buy the product. All of these offer a complete product profile that is not only informative but also impressive.

Take Context into Account- Brand communications should be relevant to the customer and the time of purchase. All past records should be collated into meaningful trends; some features allow consumers to find products similar to those that they have purchased earlier so that they are not bogged down by irrelevant entries. Information about customer location, previously bought products and interests should inform and customize the advertising the business will do in future.

Offer something different from the website- Ease of locating information is a huge perk for real time shoppers, but reiterating website details is unwise and unnecessary. An app should be

---

[5]   Latitude. "Next-gen Retail: Mobile and Beyond." Dec. 2012. Latitude.
http://files.latd.com/Latitude-Next-Gen-Retail-Study.pdf

one step ahead of the website, not necessarily in terms of the information, but in terms of interactive features.

Focus on the Essentials- Businesses should only include cardinal features on their apps. The performance of an app can be compromised if it is littered with useless items. The storage space of an app should be utilized wisely.

Provide Extra Perks- Users who engage with the business on an everyday level should be given incentives, special promotions or loyalty rewards so that they keep using the app.

In a nutshell, the advantages of having an app that best links a business to its employees, customer and potential customers is the doorway to earning profits and retaining customers. To get the largest number of users, apps should be easily available from in-store kiosks, app stores and on a cloud through desktop virtualization.

Solving Critical Mobile Challenges with Mobile Gateways

The exchange of data through mobiles is subject to many challenges. According to AT&T, mobile middleware effectively deals with these issues. "Middleware is computer software that connects software components or applications, enabling services to interact over a network. The purpose of middleware is to facilitate client/server operations, access host applications and to enable complex distributed systems".[6]

Security

---

[6] "Harnessing Mobile Middleware.", 19 Dec. 2007. RYSAVY Research. 22 Sept. 2013. http://www.rysavy.com/Articles/2007_12_Harnessing_Mobile_Middleware.pdf

In order to resolve the issue of security, according to Footnote reference, there is a need to secure data centrally and to manage policies for protecting information that is exposed via APIs. Mobility solutions should secure all data on the device ensuring there is no unauthorized access to the information.

Time

It is important to deliver and receive information in a timely manner to eliminate the possibility of a communication gap.

Multiple Devices

The app should be functional on multiple platforms and devices so that all types of users are catered for. In addition, businesses should be prepared to update their apps so that they are compatible with the technological changes of the future. However, every device differs in screen size and operating systems, thus a unified app across all platforms can be challenging to develop.

Device management

The devices that are deployed need to be carefully monitored and taken care of. The existing devices should be updated regularly so that everyone is on the same level It is also advisable that in the event of theft or loss of a device, the system should be able to erase the data or track the device to preserve enterprise integrity.

In conclusion, Enterprise Mobility solutions revamp a business' sales performance by making communication more convenient, its operational performance more automated and its employee productivity more efficient by disseminating information more evenly within the real and virtual workplace.

Extending Enterprise Identity to Mobile

According to Edu cause, "Identity management refers to the policies, processes, and technologies that establish user identities and enforce rules about access to digital resources."

Enterprise identity can be extended to mobile in various ways.

API Identity and Access Broker

API Identity means having the ability to accept different kinds of credentials for authentication. It entails that developers are allotted with different IDs and that various resource schemes are supported including federated ones like Oath and SAMLF. However, this is subject to the need to integrate with the identity structure currently in place.

Enabling Single Sign-On (SSO)

SSO allows a seamless user experience. SSO is a way to control and regulate access of various related software systems. It provides the user the convenience of logging in once and gaining entry into other systems without any retyping of passwords or user verification. Furthermore, using a mobile software development kit (SKD) will allow the development of apps for a specific platform. A cross-app SSO will ensure that users remain within the circle of communication, while they shift from one app to the other. Despite its added advantage of convenience, SSO necessitates more security at the point of initial authorization and during the use.

The system requires an SSO agent; usually a customer application installed in the mobile device. In addition, there are two components: a remote Mobile and Social Server and the applications on the device. These applications need to be authenticated with the back-end identity services via the mobile SSO agent

The SSO agent collects the attribute of the local device and send the information to the server for authorization. In the presence of the mobile SSO agent is present, user IDs are never shared with the mobile business application. "For the business app, the SSO agent asks for an access token for the resource (on behalf of the business application) and redirects the browser to the URL of the business application with the access token included in the HTTP header".[7] The tokens are authenticated before a request to target back-end services is granted (Oracle).

SSO and SAML (Security Assertion Markup Language)

SSO and SAML can be mapped to mobile friendly Oath, Open ID Connect and JSON Web tokens. SAML is used to exchange authorization data between two parties ("SAML"). "Oath is an open protocol to allow secure authorization in a simple and standard method from web, mobile and desktop applications".[8] Open ID enables users to log on to new websites or applications without having to make a username or password, i.e. they can use the identity profile created to log into several websites, which consolidates the multiple digital user identities (Eldon). "Oath talks about getting users to grant access while Open ID talks about making sure the users are really who they say they are". [9]

---

[7]   Oracle. "Oracle Mobile and Social Access Management." May 2013. Oracle Corporation 15 Sept. 2013. http://www.oracle.com/technetwork/middleware/id-mgmt/overview/mobileandsocialaccessmanagementwp-1703656.pdf

[8]   Erna Hammer. "Explaining Oath." 5 Sept. 2007. Universe, LLC. 22 Sept. 2013. http://hueniverse.com/2007/09/explaining-oauth/

[9]   Eric Eldon. "Single Sign-on Service Open ID Getting More Usage." 14 Apr. 2009. Venture Beat. 22 Sept. 2013. http://venturebeat.com/2009/04/14/single-sign-on-service-openid-getting-more-usage/

"JSON (Java Script Object Notation) is a way to store information that is organized and easy-to-read. It's both human and machine-readable and is easily parsed".[10] OAM-generated tokens are delivered in JSON format 'and' the application developer retrieves the token. The OAM policy server verifies these token and they allow access to any OAM protected resource and to numerous applications at once.

Build granular and composite policies combining Geo-location and message content

Geo-location allows one to access the position of the user, track him/her, regularly update position and use Google Maps to show user's location on the map. Such policies enable businesses and brands to engage people in a very effective and targeted way. Exchange of message content through Smartphones can be free so that there is no obstacle in communication (Gal pin).

Simplify PKI-based certificate delivery and provisioning

"A public-key infrastructure (PKI) is a system for the creation, storage, and distribution of digital certificates which are used to verify that a particular public key belongs to a certain entity".[11] It is generally used for providing authentication for applications. Operational efficiencies improve if the process of registration and deployment of certificates (Certificate Courier).

---

[10]   Dave Romero. "JSON - JavaScript Object Notation." Json.org. 22 Sept. 2013. http://www.json.com/

[11]   Mohsen Toorani and Ali Asghar Beheshti. "LPKI." 21 Nov. 2008. IEEE Xplore. 22 Sept. 2013. http://ieeexplore.ieee.org/stamp/stamp.jsp?tp=

Federated Identity Management

There is a need for uniformity in the practices, policies and rule for security followed by everyone in the organization. This implies powerful adaptive federation and authorization capabilities to enable security in a timely manner according to the needs of each user.

Secure Mobile Access

The security of the information should not be compromised if a different device is used to connect to it. Therefore, the system has to be failsafe.

API Security

The IT staff should be able Identify who is accessing APIs. "The WS-Trust Security Token Service and Oath Authorization Server Allows developers to include identity information in their SOAP or REST-based API calls using open standards." [12]

Social Identity Integration

Allowing users to collaborate all their identities on social media is another way to increase the rate of app adoption and usage. The user's identity on Google, Yahoo, Twitter, Facebook, Windows Live or Twitter can be organized to give a business more insight.

Automated User Provisioning

This facility allows enterprise cloud services to be updated automatically. Use "SCIM (System for Cross-Domain Identity Management), a lightweight, automated standard for inbound and

---

[12] "Identity Management Solutions." Ping Identity Corporation. 14 Sept. 2013. https://www.pingidentity.com/products/pingfederate/

outbound to corporate directories, replacing costly proprietary or manual provisioning methods".[13]

API Developer Portal

Businesses need a way to control and track the broader operational character of how APIs are exposed to different partners and developers, through policy characteristics like metering, SLA, availability and performance.

Developer Portal

- Plans- API Explorer- Developers should be able to discern how an API works.
- API Documents
- Quotes
- Ranking
- Analytics- the performance of the application should be recorded and analyzed so that the organizations would know which opportunities to capitalize and which problems to solve.
- Forums- Stimulate developer communities

API Integration Gateway

When enterprises move between different environments, locations or databases, they need to make sure that API updates do not break.

- Transformation
- Routing
- Traffic Control
- Composition

---

[13]   "Identity Management Solutions." Ping Identity Corporation. 14 Sept. 2013. https://www.pingidentity.com/products/pingfederate/

- Throttling
- Prioritization
- Catching
- Security—providing critical protection needed between un-trusted and trusted zones.

API Service Manager

The API should work flawlessly, with absolutely no downtime, which could affect the end user. Enterprises simply cannot risk this, as their entire reputation could be jeopardized by any malfunction.

- Health Tracking
- Performance- Collect usage metrics (for insight into performance and for billing purposes).
- Global Staging
- Configuration Migration
- Workflow
- Patch Management
- Policy Migration
- Reporting

API Management Solutions

The management solution needs to be compatible with existing technology and investments from companies such as Oracle, IBM, CA and RSA. However, API security does not simply stop at controlling access.

APIs act as a software-window to your enterprise's data. Hence, an API management solution must give the security officer very precise and comprehensive controls regarding what information is released, what is kept confidential and how they can safely transfer information from one point to another. Finally, API security relies on not only the API but also the functionality it provides. A good

API management system solution should provide its operator with an entire arsenal to protect against threats.

Developer Enablement and Community Building

Enterprises should focus on bringing developers on board, providing training and assistance in order for them to get as much as possible out of the exposed APIs.

- API Monetization –APIs also represent a novel opportunity to generate revenue.
- Solution Security – Businesses require comprehensive and powerful API solutions that satisfy a wide range of security needs, from protection from penetration to PCI compliance to FIPS to HSM support and so on.

Solution Manageability – Businesses have development and production facilities that can span several locations, databases and data-clouds.

- Solution Reliability – Businesses cannot afford any down time. This the reason IT managers, web managers and software architects are requires to have provided to them, precise information to allow them to better select an API management solution.

The aim is to integrate the new with the existing infrastructure; achieving adaptability and seamless integration. This should include the ability to allow support for different types of access tokens as well as the ability to not have to alter the code while moving from one kind of developer API key to another.

API Governance in itself is a very broad term. It is used to refer to a variety of management processes and visibility requirements. It explains the circumstances under which an API is exposed to customers. Governance not only includes security and life cycle concepts, it also covers SLA, monitoring and reporting requirements. A good API is all about adaptability. Hence, APIs should adapt to the enterprise in question as seamlessly as possible.

A developer portal provides the developer with the options to register for an account, request access keys and find out what APIs are available. A good API developer portal should:

- Support different types of developers
- Provide self-service capabilities
- Allow developers to monitor API usage and key performance figures
- Share information with other developers about the most efficient methodologies

A 'one for all' API approach is unrealistic, since different enterprises have different setups, requirements and objectives. An API portal has to be as customized and integrated as API security, life cycle and governance framework. Therefore, many enterprises prefer a decomposable API portal. This usually means a basic portal that can later be customized by the enterprise to perfectly match their requirements. It could also mean that different features of an API portal could be used separately while complementing the existing developer portal.

API monetization refers to developer enablement. Many enterprises prefer to allow free access to their Web and mobile APIs, since this increases adoption, many others instead opt for a pay-per-use sort of a solution. This is mostly for higher tiers of access. Overall, there is no one solution for everyone.

Translate and Orchestrate Data

APIs for business should be able to manage data from various sources. This allows automation of Business Processes Management (BPM). It is essentially an executable code to support workflow for a business process.

This 'Orchestration' of data from outside sources is done by using various receive and send ports. Here are some common terms and their definitions:

- Orchestration: Business interaction that is captured through shapes
- Transformation: create a link between source and target
- Message Correlation: receiving and forwarding data in the same orchestration instance.
- Orchestration Designer: a tool to design orchestrations.
- Promotion: these enable various server components to access key items of data.
- Distinguished fields: these are only accessible from within orchestrations. They do not require the creation of a corresponding property schema.
- Dehydration: an instance is dehydrated if the orchestration determines that it has waited beyond a certain threshold.
- Rehydration: Once an orchestration receives a message it loads the saved orchestration and runs from where it left off.

Orchestration Engine

- Creates and executes instances of orchestrations
- Maintains and saves an orchestration to allow it to start from that point when it needs to later on
- Optimizes running orchestrations to maximize efficiency and effectiveness
- Reliably shuts down the system as well as a recovery point

Orchestration: Steps to develop

- Create the schemas
- Create and define the shapes to signify the business process

- Create new message instances
- Create and assign variables to manage the data utilized in the orchestration
- Create the ports which receive and send messages
- Change the send and receive shapes for the ports
- Test the orchestration by creating and deploying to the GAC

Orchestration: Message routing

Message routing occurs once a message is received, an instance is activated or an existing orchestration is hydrated. Whenever an administrator needs resources through the resource orchestrator, he has the option to receive them automatically. Once he defines the pattern, the orchestration automatically assembles it for him, which can lead to much higher productivity.

Managing the Scale of Enterprise Mobility

Mobility Management refers to managing the life cycles of applications, data, devices and APIs. It is also used to refer to building applications efficiently and effectively, connecting to data mobility management systems and supporting multiple devices. As the business expands, it needs to adapt to the new scale, get big data interaction and perform cloud integration.

- How to scale development: Choose from a variety of options to develop your mobile application.
- How to scale deployment: App providers are the source and they manage the development life cycle, protect Mobile IT, deploy apps at a scale and support remote users; only then, it should be deployed to the final user. Apps should be rolled out at a scale to achieve enterprise mobility ROI (Return on Investment). Generally, apps are adopted through network effect; everyone in the chain uses the app.

- For success, assimilate the Self-Serve solution where users can sign themselves and maximize user satisfaction by providing familiar, easily discoverable apps that can be customized and used on multiple devices. Ensuring IT comfort and compliance is achieved by having a unified admin portal, secure apps, and cloud based/scalable system.

Enterprise mobility solutions should be scaled to the business demand, which is made possible by:

- Management- Track and manage the mobile device life cycle
- Provisioning- simplify and expedite the deployment and decommissioning of mobile devices and approved applications, including full and select remote data wipe.
- Security- transparently secure mobile devices, their corporate data, and the IT network they access. For iOS5 mobile devices this includes requiring trusted certificates to avoid potentially malicious web sites
- Support- Minimize IT support costs and maximize user productivity
- Audit- Support enterprise IT and policy compliance reporting requirements

Mobility Management systems allow operations to improve the efficiency and effectiveness of deploying mobility solutions. It does this by converting complex installations to efficient and reliable solutions. These are much easier to manage and deploy. Other benefits include faster development, deployment and problem resolution.

Due to standardized application architectures, better operating systems and wireless LAN mobility solutions are now scalable and. In addition, mobility deployments that are centrally managed and regulated have reduced complexity and risk than previously existed. The need for security and focus in executing a mobility plan requires a single-minded approach for support, control and development of mobile devices. It also allows IT staff to monitor costs and oversee the performance of the system in real-time, so that the contingent problems and progress can be identified.

Motorola Solutions states that to overcome the challenges of "device rollouts, mixed application deployment, manual updating of applications, isolating problems, reactive monitoring, intermittent connectivity and ensuring security, the Mobility Services Platform was introduced"[14]. It takes a unified approach to supervise and govern mobility deployments.

Centralize Cloud Connectivity

An organization as a single unit cannot manage vast amounts of network or user traffic. Cloud solutions provide a way for this burden to be spread out. Cloud Connectivity "allows companies to free up private bandwidth usage while maintaining centralized control of web traffic by utilizing local lines at remote locations."[15] It permits businesses to centrally manage corporate-wide web access policies by deflecting the common issues and costs related to increasing bandwidth. It can help store business data and host applications somewhere other than the premises of the enterprise. It stretches the capacity of existing IT infrastructure without any physical acquisition of hardware and servers or spending on training of personnel. It is a method whereby storage, computing, security, filtering 'and' authorization can be outsourced.

[14] Enterprise Requirements Planning — Scaling Enterprise Mobility Deployments." Motorola Inc. Web. 15 Sept. 2013. http://www.motorolasolutions.com/web/Business/Products/Wireless%20LAN%20Devices/Wireless%20Switches/WS%202000/_Documents/Static%20Files/ab-35_tb-erp_0305_fnl_new.pdf

[15] "Cloud Connect." FatPipe Networks. Fatpipe Networks. 14 Sept. 2013. http://www.fatpipeinc.com/cloud/index.php

Originally, IT departments tapped into cloud solutions individually, but now new integrative forms are emerging that offer unified solutions by sharing resources and reducing costs. Instead of procuring software for each new system, the firm can load one application that contains programs, which can be used by everyone. For business emails and document sharing, such a system is very convenient because it uses minimal space on the mobile device itself, since all emails and the data they contain will be stored on the server in the cloud. Cloud connectivity gathers user requests from remote locations and forwards it to a designated central location to grant web-access authorization. Once authorized, the user's requests are catered to by a local line in his/her location instead of the private lines.

Combining cloud solution with mobile devices is important because the market is subject to upsurges in users and rapid shifts in technology. There is an increase in the types of devices used; smartphones, tablets, laptops. Likewise, the number of users and applications have also multiplied. To ensure that the users do not experience an interruption in their service, it is imperative to restructure and synchronize data storage by shifting away from old databases (legacy systems) and to a flexible system of back-end storage that can accommodate an increase in demand. Ultimately, the emphasis should be on user experience, not business storage issues. The cloud system at least absolves the business of this worry.

Below is the comparison of a cloud and a mobile device:

Cloud	Phone
Located somewhere far	Geo-located
Scalable Processing	Constrained Processing
Power freely available	Battery problems
Large, fast storage	Small, slow storage
Always connected	Intermittently Connected
Scaling cost	Free

Basically, what drives the adoption of cloud computing is economics; an illusion of infinite resources 'and' the billing can be hourly or per unit of memory, which results in vast economies of scale in the data Centers. The technological impetus for adoption of cloud computing includes widespread broadband Internet and maturity in virtualization. The advantages of using a cloud include agility, scalability, security, maintenance and metering. However, it is subject to legal challenges, privacy and sustainability issues.

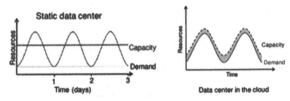

The under-provisioning of resources can also cause problems of poor quality service. The data center in the cloud caters to the demand very closely (as shown above).

The workload can be distributed across a private cloud, an cloud, a collocated datacenter, a cloud platform 'used' a service provider and/ or the cloud application. An enterprise is likely to centralize all of these into a virtual private cloud and run it as a centrally managed single unified datacenter. "Virtual Private Clouds will span applications that will be deployed on-premise, off-premise, in the cloud, and in application-specific public clouds."[16] They will be secure, centrally managed, with identity and single sign-on federated across these systems.

[16]   Anand Deshpande. "Cloud and Mobility (slideshare)." 21 Sept. 2011. SlideShare Inc. 14 Sept. 2013. http://www.slideshare.net/anandesh/cloud-and-mobility-slideshare

The management layer of the virtual private cloud must manage identity by allowing single sign-on for applications. It should also host the Business' app store containing every app a business offers its users. Handling varying loads, billing accuracy, supervising multiple systems that are running within a cloud and designating workloads to the most places are additional tasks of the management layer.

Centralizing the cloud solution is necessary in order to allow a business to amass large amounts of data about its users for the purpose of efficient and effective marketing, advertising and customer retention.

There is, however, the issue of security. Often time's users of an app are concerned about sharing both private and public information with a business or organization for fear of unauthorized use of that information.

One example of this is Google[17]; it accumulates vast amounts of data about the user's preferences, their social demographics, their activities, places they have been to and so on. Nevertheless, the quality of service provided greatly 'because of this'. If the mobility solution uses centralized cloud connectivity, then it can evaluate and analyze the metrics of user behavior. This is a cost-effective form of research too.

Security Solutions

Mobility solutions are all about the transfer of information conveniently. Securing data during these transmissions is vital for the integrity of the business. The two aspects of access and

---

[17] Primavera De Filippi and Smari McCarthy. "Cloud Computing: Legal Issues in Centralized Architectures (July 11, 2011). VII International Conference on Internet, Law and Politics, 2011. Available at SSRN: http://ssrn.com/abstract=2097146

protection need to be balanced sagaciously; tipping over to either side's favor is sure to prove problematic for that business in the future. With the advent of mobile systems, organizations have become more exposed to external threats. Data is no longer stored in a document that can be locked in a safe; 'meaning that' the flow of virtual information and operation is nearly impossible, but highly essential.

The Life-cycle approach to a Mobile security strategy

The Lifecycle of a mobile device is defined so that security policies can be implemented at each stage.

Provision phase —is when a business starts using a device. At this stage, configuring the device with the right security software will ensure it functions properly for some time. "Device initialization could include separating business and personal files, installing antivirus software, providing business apps, setting up password protection and configuring network access".

Production phase — a configured device is ready to be used by the business. Obviously, it requires consistent updating with modern software so it does not fall prey to newfangled malware.

Decommission phase —the device is no longer used by the business due to being lost, stolen or replaced. It entails deleting all business data, applications, and other content so that it does not fall into the wrong hands.

TELUS delineates its security roadmap[18] :

---

[18] "Security." Telus. 14 Sept. 2013.
http://telus.com/en_CA/National/products/Medium_ And_Large_Business/Security/natMlbSecurity.html

1. Application Security:

    a.   Software Security

        i.   Secure Development Model Software: There is a need to provide licensable intellectual property, developer training, and consulting services that greatly accelerate a software development organization's adoption of best practices for secure development, across all stages of the software development lifecycle.

        ii.   Source Code Reviews: Detecting as well as eliminating security vulnerabilities at the source code level has been costly and time consuming, increasing the risk that hidden issues will go undetected. Training developers in code security deficiencies and remediation can overcome this obstacle.

        iii. Web Application Testing: This is highly scalable and cost-effective.

        iv. Software Security Awareness Training

    b.   Web Application Firewall: ensure full protection so that every aberration from acceptable use of the application will be detected and blocked. It should be manageable in the sense that customers' requests conforming to permitted use of applications is never blocked. The firewall should not impose a performance overhead. Digitally signed records should be saved and protected again deletion or changes.

2. Data Security:

    a.   Security Information– using technology that allows the business to collect logs from the entire critical infrastructure.

    b.   Log Management

    c.   Outbound Content Control –hard disk encryption and content monitoring solutions

    d.   Managed Secure Authentication –Other than a password, provide a second layer of security to protect from unwanted access.

    e.   Email Protection –Avoid potential downtime or slowdowns that result from viruses or spam. Scan and filter all incoming and outgoing e-mail while protecting the privacy of your communications

    f.   Web Content Security—filter and report the Web activity.

    g.   Desktop Backup – to protect data if the device is lost or stolen.

9. Digital Forensics—Analysts can help track user and business information and detect cybercrime.

4. Governance, Risk and Compliance

    a.   Enterprise Security Program

    b.   Threat & Risk Assessment --refine the information and privacy risk management strategy

    c.   Policy Consulting --develop security policies and procedures based on: business aims, existing abilities, current state of security, existing threats and regulatory requirements.

    d.   Privacy Consulting

5. Infrastructure Security

    a.   Secure Network Architecture

    b.   Intrusion Prevention – Use your network to identify abnormalities and potential damaging activity and block it.

c. Managed Firewall and VPN --Secure the traffic flowing in and out of your network with Managed Firewall and Virtual Private Network.

d. Managed Virus Protection –protect the Internet gateway, e-mail servers, file and print servers from virus afflictions

e. Infrastructure Security Testing

f. Vulnerability Scanning

10. Mobile Security- within the device the following features ensure security.

a. Antivirus

b. Anti-spam

c. Personal firewall

d. Loss/theft protection (tracker)

e. Backup and restore

f. Application monitoring and control

7. People- train staff and employees for securing data

11. Threat and Vulnerability Research—find out areas of high-risk and fortify digital barricades to prevent intrusion ("Security")

Remote Wipe Technology and its Benefits

The ability to remotely wipe any managed device is indispensable to many enterprise mobile security policies, and it is vital to preventing sensitive corporate data from being compromised. In any case, employees/users need to be aware of the Option to remove all data and prevent anyone from accessing it in case of theft.

Remote data wipe options depend on the type of device, the operating system and the security application in place every business should have a protocol for remote wipe options and a clear set of guidelines as to when to apply them. This can then be implemented onto each device carried by members of their workforce.

Remote Wipe Options

Factory reset: This deletes all device settings, data and applications from the memory storage. This is an easy way to obliterate all information and there are ways to retrieve it if you need to do so.

This is an easy way to obliterate all information and there are ways to retrieve it if you need to do so.

- **Business device wipe:** all information that is related to the business, including business email, apps, account settings and other data is deleted. It does not, however, erase personal data or other apps on the device installed by the user.
- **Secure container removal:** Only a data container with files or folders of the Business is removed. This option makes those documents unreadable by deleting the keys used to encrypt them.
- **Secure app removal:** A specific business app is deleted along with its data, 'but' rest of the information and apps are not affected.

With more and more business transactions on the go, cell phones are now the weakest security link in many organizations. A phone may contain sensitive private or business data like company documents or files somewhere in the cloud.

The functionality of a remote app is contingent upon its specific operating system version and any third-party mobile device management (MDM) software installed on the device. Removing programs, deleting specific data or overwriting files are ways that remote wipe provides security. This is different from local wipe- a facility that deletes data if an unknown person is making multiple attempts to login or if the user moves out of a specified location. Remote wipe technology ensures that data does not fall into the wrong hands and that business can protect its integrity and information from prying eyes.

Why is Password Protection an Essential Mobility Tool?

Password protection is perhaps the most convenient and easiest way to protect invaluable digital resources. However, its ubiquity has undermined its efficiency as a tool for security. Handling more than one password is counterproductive. Passwords enable the system to identify the user and to allow access.

Password Protection as a Business Solution

Firm and strong user authentications are the cornerstone of an uninfringeable security system; thus enabling password protection is important for protecting data and identities. This increases productivity because users can securely access business applications and their accounts. Additionally, they can conduct transactions without the fear of losing their personal and digital identities ("Password Protection"). Simply said, if users have trust in the system of their device they will be confidently mobile. If they are unsure about accessing information from anywhere then mobility will be adversely affected.

Today's mobility is complicated by workers using their own devices for both work and personal purposes Security is further complicated by the increasing diversity of mobile devices in the workplace and a proliferation of mobile applications that make security management an arduous task.

Safe Net offers the ultimate power and flexibility to implement effective password protection and secure access to digital business resources.

Case Study

For example, needed a strong, foolproof security solution, which could protect their high-risk transactions. They wanted a solution that could assure the security of their private keys in multiple applications. The business also needed a multi-factor authentication solution to verify identities of external and internal users. Working closely with Safe Net, Prom invest bank designed a security infrastructure that incorporated key protection, authentication tokens to authorize end users, and a centralized authentication management.

The Safe Net betoken 5100 was used as a portable authenticator that uses smart card technology to provide secure access. It stores user information in the chip and while making a transaction the user is supposed to provide the electronic token and the original password, so that the security layer is (Prom invest bank Case Study).

Mobile Web Apps as a Viable Solution

The mobile web refers to the use of browser-based Internet services, through a mobile device. Developing hosted, mobile web applications can offer an attractive and viable solution that can overcome the fragmentation and deliver cross-platform services.

In order to effectively implement a mobility solution, there is a need for users to adopt the application and it needs to be functional across different platforms. Every platform has different SDKs (Software Development Kits) and programming language and the enterprise has to spend the money to overcome each of these obstacles in order to develop a unified app. The need for an app is unquestionable, but the intelligent approach would be to opt for a more viable option like a Mobile Web App.

Mobile web applications can overcome fragmentation in mobile operating systems and app stores by facilitating the development of apps that run on multiple platforms and devices, using web technologies such as HTML, CSS, and JavaScript. These apps can then be hosted on the existing web server and accessed through the web browser on the device. Developing and delivering web apps automatically eliminates the need to pay web developers for coding, or the app stores for publicizing the business app. The building and deployment is through the web, so many clients can be acquired easily. The support for HTML5 and modern web standards is increasing the viability of adopting mobile web apps.

Key Features of a Mobile Web App[19]

- Video without Plugins: web developers can now include video within their pages without the need for embedding it in a plugin like Flash.
- Local Storage: the ability to create applications that store their data on the user's phone rather than the servers. Application features like calendars, personal notes etc. can be used offline and even the personalization of the app can be done without logging in.
- Offline Applications: offline capability of app brings it closer to the native experience as your key interface features – buttons, images, styles, scripts– can be functional even if the user has a poor internet connection. Some files can be locally saved on the device so that there is not a blank screen if internet disconnects.

---

[19]   Mark Power. "Mobile Web Apps." Mar. 2011. Jisc Cetis. 13 Sept. 2013. http://wiki.cetis.ac.uk/images/7/76/Mobile_Web_Apps.pdf

- Geolocation: it can enhance the user's interaction with the business service by pinpointing their exact position; this is particularly beneficial when businesses can suggest nearby stores or services to the user.

- Multi-threaded JavaScript: provides applications with the ability to use scripts that run in the background without interacting with users. These can be used for long-running tasks or functions that require a lot of computation e.g. scientific calculations.

- Easy Form Handling: HTML5 brings new form types that are recognized by the browsers and formatted accordingly; presenting the user with the keyboard they need, not long and lengthy JavaScript (Power).

Some of the advantages of using a Mobile Web App are:

- The advantage of being able to use the app on multiple devices – iPhone, Android, Blackberry, etc. through the mobile web browser

- HTML, CSS & JavaScript will get help in developing the app. Knowledge of complicated coding languages is not essential.

- Instead of using the software development kit (SDK) of a specific platform, the authoring tools preferred can be used.

- There is no need to get apps approved or to spend more since the app can be hosted on the organizations' web server.

- The app can be updated very easily without having to deal with an app store. Therefore users will always have the latest product

Some of the disadvantages are:

- The app can't fully access all the hardware features of the device
- It may lack some of the sophistication of a native app when it comes to user interface and experience
- It can have a slower performance than native apps

Mobile web apps are suitable for smaller business or those who need a simple app to cater to their mobility needs.

CHAPTER **8**

Scope of Enterprise Mobility Management

Enterprise Mobility Management is important to firms that have global integration on their agenda. It is an implementation meant for the entire organization and attains maximum efficiency when fully integrated and operating in all business sectors.

The Many Dimensions of Enterprise Mobility Management

When dealing with enterprise mobility management, businesses need to understand the vast scope of mobility software. Once mobilization is ensured among different areas of a business, the managing team must remember it is not merely the people under question. Apart from the employees being made mobile, a business will come to understand the need for vigilance in a number of areas including: financial records, customer interaction and end-user activity. There has been an uptick in interest in machine-to-machine communications driven by businesses able to justify the cost of mobile services through the benefits they offer in terms of improving business processes and creating competitive advantages. Many Businesses have already successfully implemented mobile field sales force automation through solutions such as Salesforce.com. These same businesses should now consider opening up their other enterprise resource planning systems and thinking seriously about the benefits of real time tracking of deliveries, technicians, and goods.

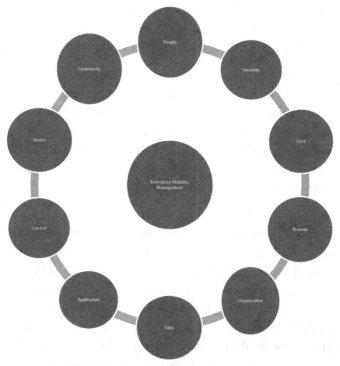

In the growing world economy every business needs to implement enterprise mobility into their operation and manage it in a way to integrate all business functions, departments and devices of all types. The many dimensions essential to enterprise mobility management can be loosely defined as follows:

People

The employees and consumers should be seen as the key players in any mobility management system and treated as such. In many cases, however, this "people" dimension of enterprise mobility management can include entities or devices other than humans.

These counterparts are the machinery or devices that constantly perform a task for the company in the form of analysis and response of certain conditions or requirements of the working process. These devices may or may not require human supervision, but will operate in the absence of human intervention.

For a business to attain proper of its mobile platforms and devices, it must first understand the characters and their specific roles in the enterprise. Once the people are given their mobility devices, they must know how to operate them and make relevant real time decisions. The People must be offered adequate training and must preferably have experience in the field of enterprise mobility management prior to the implementation.

The People should be provided with devices to use to make work easier and more mobile. : A business must also remember that safety and security of operations should be paramount when organizing and carrying out their mobility system. The fact that the device may contain applications or software that might pose potential harm to the Business must be considered when handing out mobility devices. It is advisable for businesses to manage applications on their system, implement software to separate work related activities, and safeguard Business-related apps from non-work related activities. With the wide array of device brands available in the corporate market, the enterprise must consider the safety regulations and applications it must purchase for security reasons.

Location

Simply put, to enable effective enterprise mobility management, the workplace or "Location" (the place where people do their work when at the office or outside) dimension must be considered. This will never necessarily be one specific location because mobility entails work processes spanning different geographical locations.

Some however, might be unsafe to work in and some might be considered to improve employee productivity due to their characteristics.

Other than this, the location dimension of enterprise mobility management is important because it affects the communication between employees and customers (or suppliers), the access to relevant devices or workstations to enhance time efficient working and can also affect the reach of the company network to the employee who is in possession of the device.

Process

This will include the work specifications of the people. The process dimension includes the tasks and the order in which these tasks are performed. While carrying out an operation to ensure effective enterprise mobility management, a firm must know what process is being followed; whether it is a routine process, an on the spot decision by the worker, or a planned process for a specific time.

The process will be the amalgamation of different tasks that the Business and its workers put into practice and ensure regularity in so that the corporate aims are not compromised. For enterprise mobility goals to be met, the process must be clearly defined and indiscreet, following a proper set pattern and routine to be manageable and reportable. However, everything may not always work as a typical assembly line and a systematic way may not be followed in carrying out processes. In this case, the process dimension will not be an easy one to cover through enterprise mobility management techniques, and individuals will have to be employed to monitor these Processes.

In the process, one very important factor is technology. Technological innovations, the skill set of the workforce, the knowledge base of the enterprise, capital available for use and other factors also affect the processes. These factors must be known to the management team in order for them to understand the dimension of the process in enterprise mobility and its management techniques, as well as how to maximize returns to hefty investments. The IT department is of great importance to a firm's enterprise mobility management system and helps integrate core business processes to meet mobility requirements, be it through implementing small changes in processes, or changing complex sectors of activity in the Organization and People.

Processes previously had to be carried out through stringent and boring practices in the form of a production line, where all workers needed to be present at the workplace to ensure timely operation once the previous step was completed. With enterprise mobility software 'and' applications and devices, this old and mundane work process has been revolutionized and can be carried out more efficiently, while the people are present at the location where a decision has to be implemented and make Sound decisions to the prevailing conditions.

When smartphones and portable computers are added to the equation of process, the entire setup of activity faces a change. The change comes about in a way where processes can be decentralized and more diverse even though the output at the end of the process may be no different from what was initially planned. A more decentralized process structure will give workers the feeling of empowerment and help them feel they have achieved more than they had before. The empowerment comes in the form of more autonomy in the hands of the people in terms of small-scale decision making for their own productivity enhancement. However, the lack of training and experience among people can Problems in enterprise mobility management. If workers do not know how to operate their devices and Make multiple and/or continuous errors in their work, the management aspect will fail, as the process well be disrupted.

Organization

The organization will represent the person or entity that implements mobility in the Business, and must deploy devices and applications to ensure correct working of them. This entity is on whose behalf the workings are carried out in a way that is not geographically constrained. The organization can be anything, from a client to a firm or a single individual.

The major interaction in business processes is between the people and the organization. To ensure safe enterprise mobility and ensure effective management, legal methods should define the chain of command and the corporate relationship between the organization and its people. The legal papers should be signed by both parties. They should clearly and explicitly explain the job duties, terms of use of mobile devices and code of conduct in the event of theft or misplacement of the mobility devices in the possession of the employee.

The responsibilities of the organization in enterprise mobility must be identified. Is the organization meant to oversee the work done by people D or are they supposed to report to the organization? This can be one case, while another can be that the organization has no part in overseeing or controlling the actions and decisions of the people regarding use of mobile devices. A change in the political structure of the firm may also be needed. This can be the case if the nature of the Business was autocratic and a more laissez faire approach needs to be adopted for mobility to be ensured.

The organization has a key role to play in the implementation of mobile platforms, because the process revolutionizes organization–people relationships. The people must be aware of the corporate aim of the organization and its goals for activity to be able to make and quick mobile decisions. For such decisions to be effective, the organization must make sure it is in sync and effectively communicates with its people.

On the other hand, if this relationship is not permanent or and spans over a short time, this take on this specific dimension will be inaccurate. It can affect the kind of tasks being carried out on the mobile device.

Data

The private information about a company and its use is one of the highest value products for a firm. The protection of a company's Data is of vital importance. The information will have to be protected from unwarranted use by the people or misuse of devices Connecting to unsafe networks and operating the mobility device in Unsafe

Areas can also pose a risk to the company's data.

The data of a business is relevant to the employee should be sent to him in a timely manner to ensure effective and efficient working. To make this happen, a fully integrated enterprise mobility revolution will have to be brought about in the entire organization.

Application

The application is the software that runs the devices in a mobility system. It can also be the program that the people use on their devices to simplify and enhance work related tasks. The Application helps simplify the People's tasks, and helps them carry out actions upon predefined Processes, value judgments or routine exercises. The Application dimension is easy to manage, implement and learn to use by everyone within the business, and requires minimal management and work by specialized teams. Extremely high levels of sophistication in Application on mobile devices are available to enterprises range from simple presentation creating Application to highly sophisticated SAP integrated application for official use.

A fun fact to know about some Application is that they can store data in the cloud. Cloud computing has been around for a very long while and the integration of this into mobility devices is a breakthrough for Businesses wishing to understand the dimensions of enterprise mobility. Data stored in the cloud can be accessed through mobility devices using a simple connection to the internet. This ensures security of the work done on a device, as no unprotected transfer of data is carried out and the worker has his own constantly update source of information from the firm.

Control

The organization has to make sure it exercises a certain level of "Control" over its D the D dimension of enterprise mobility management means that the organization must be able to use mobility as a tool to manage and oversee the decisions being taken by its People. The Control dimension can be in the hands of Management, or a special team that manages and overlooks responsibilities and decisions by People in areas where they are not experienced or where Management feels guidance and control over actions is necessary.

In most cases, longevity on the part of an employee enhances his/her relationship with the organization is likely to be In turn, this usually results in less control being executed by the organization due to a mutual trust and loyalty between organization and these people.

In spite of trust and loyalty, a business still needs to maintain a certain standard and amount of control. The company must know where and how the device will be used when on the go. The type of work someone is doing also works to determent the amount of control which needs to be exercised over the employee and the mobility device. The organization must also know when, where and how to exercise any necessary control over the employee's actions with a device. There is no room for loopholes in the network, app design or device configuration which could allow an employee to work outside the perimeters controlled by the organization.

Apart from technological problems, the danger to safety is essential due to the fact that the people can sometimes be operating in a manner independent of the firm's control. This normally happens in the event that an employee is put in the position of making certain decisions themselves. In such cases, controls can be placed on the security of the device against unlawful access and theft of corporate data, but the efficiency cannot be guaranteed as in this day and age, hacking techniques are rapidly gaining sophistication that computing managers are trying hard to manage and beat. Moreover, if the control the organization has over the work that is being done by the employee in a far off area is weak, the organization may find it near impossible to oversee or protect the transaction and decision.

Community

When any task is being carried out, there are several entities, employees and subordinates who have some level of involvement in the process, in the form of suppliers, supervisors and customers. This is what we call the community that exists in an organization and in some cases the community cannot be controlled by the organization or the people who carry out the tasks.

In any mobility system, a clear definition as to the role of the community in regards to the individual usage of a mobile device. The amount of information provided to the community, their involvement in processes being carried out and the amount of communication expected between the community and an employee are issues that need to be decided upon to make sure the mobility system remains beneficial and cost-effective.

Device

Devices are the mobility enabling hardware that is made available to the people to make their work less subject to geographical constraints. The device can be anything from a smartphone to a laptop or even the pocket projectors as discussed in Chapter 4. It can be made available to the Organization as a whole, or to specific people, pertaining to their job details and descriptions.

The Device does not include the technological and communication aid that makes it a contributor to mobility. Many other investments will have to be made by the firm to ensure that the device offers the desired result to the Business. Other than this, several additions in the form of extension devices and more equipment will make cost management a matter of concern as the cost for deploying accessories to employees is rather high.

The Device may not always come fully integrated with the company's work dealing etc. and several changes in the interface, connectivity and database recognition might have to be made by the IT department; thus adding even more to the cost of the system.

The device should be one that is at least somewhat familiar to the people. It needs to be one they are familiar with using or can be trained to use with little effort, in order for them to use it to their maximum advantage. Even more important is the fact that everyone needs to understand why the mobility is being put into action. What role or roles does it have in enhancing efficiency and productivity? Everyone involved needs to know the answers to these questions and how to do their part in making that happen.

Cost

Easier said than done, the cost of a mobility system is two-dimensional for any business. Officials are paid for their hours of work; the time they spend creating and managing enterprise mobility systems comes at a tradeoff of cost and output. The Cost also covers the payments made for devices and software that are implemented at the firm, and training (if required) to be imparted to the people of the firm.

Costs can be relative and absolute for both the business and for enterprise mobility, 'but' accounting and management of costs is necessary. Cost benefit analyses, cost structuring, ROIs and all other aspects are developed and managed under costing. . The cost may not be restricted to just the Business (and end sentence here) It may be possible that end customers or suppliers would have to incur extra costs in communicating and interacting with the business Upon identification of the several dimensions of enterprise mobility management, firms will be more capable of attaining operational efficiency and will be able to minimize risks of going mobile at the enterprise level.

Device Management

Device management falls directly under enterprise mobility management as it involves the use of software to ensure security, management and constant watching of devices that are deployed to the people in the firm.

Considering the sphere of management of mobile devices, specialized software is used to monitor and analyze the working of different people, organizations and clients. The management process will mainly include the transfer of data through internet and intranet to devices which are provided to the firm and its employees and even to the devices that the employees bring if the business has a BYOD policy underway. The aim of device management is not complicated and is a rather simple one. What it aims to do is ensure secure transmission of data to and from stakeholders, while keeping the costs to the 'as low as possible'. Device management involves a team of people (or applications software) to keep track of the configuration of all devices in the company's network to maximize security.

As the BYOD trend and use of consumer devices for corporate use is taking over in enterprise operations, device management is of utmost importance. This is because the BYOD mobile devices in a business mean extra steps and precautions to be taken for device management to ensure highest security of operations. Many providers in the market help businesses and cell phone developers test their devices and earn hefty amounts of money through providing prerelease device management. This can also apply to applications or applications solely meant for the use of the stakeholders of that organization.

With cloud computing and wireless computing being the norm, it is essential for programming of devices to be possible in a business. The programming software is extremely helpful in configuring a single device or multiple devices in a go. This is great for company security, as the protection of data is ensured. Upon reporting a theft of a device to the firm, an employee can be sure of immediate deletion of all possible data from the device. This can be done through a binary SMS being sent to the device (or mobile) that the company wishes to configure or delete data from.

Different companies and providers ensure the provision of device management services to businesses. For example, popular device management software Citrix Xin Mobile helps companies achieve enterprise mobility and simplifies device management. Such products have USPs in the fields of high security and flexibility while the people use the device as they please at work and home.

Device management ranges from Samsung KNOX packages to Blackberry's own software which help data protection to the ultimate degree, by separating the user's personal information from official information throughout the user's employment at the firm. This ensures that no unprotected access is allowed to valuable company data. The use of IT to ensure device management and security is quintessential and aids the ease with which devices can be deployed with no restraints upon the user's need and personal requirements and entertainment from the device.

Supporting and Cooperating with the APIs (Application Programming Interfaces)

API is an acronym for Application Programming Interface and its major use in in the world of business involves enabling specialized development of applications through defined processes, guides and protocols. The advantage of APIs is that it helps enable programmers to build custom and applications intended to be in many or all facets of a business.

It is an interface which basically allows programmers and users to interact with other programs. It allows developers and companies outside your sphere to integrate your information or application detailing in their own developments.

APIs can be used by professional programmers or simple device users with little programming knowledge. This is made possible through the similar interfaces of applications that have a common API; 'meaning' to these new software and programs is not difficult for users. If a programmer requires use of another program to carry out a task, he uses an API (certain standard procedures and steps that have been defined for all Insert) since every application and program makes use of APIs, the programmers have a way to make standardized requests for specialized services from any program they desire. The API provides a way to ask for certain services to be carried out by a program and improves the process of opening and gaining access to programs and applications.

In the business world, the use of APIs is of great importance and the SAP API provided to enterprises is BAPI (Business Application Programming Interface). BAPIs serve people by enabling access to SAP functions across formal, stable and dialog-free interfaces. Customers and software partners may develop external applications to access these interfaces. The object types defined as API methods of SAP Business Object Types enable object-based communication when used within the Business Framework.

In companies, BAPIs and the related business objects allow object orientation in central information processing to the extent of implementing non-SAP components, and reusing existing data and functions. BAPIs may be used by applications to gain direct access to the application layer of the R/3 System, and may also make use of the R/3 business logic through applications running as clients. Through BAPIs, the client is provided with an object-oriented view of the application objects without the required implementation details. System-wide business processes are mapped and implemented by defining scenarios in the development of BAPIs.

In the past few years, the use of mobile devices to facilitate business enterprise working has risen greatly. Due to this, workers and employers need to access data on the go and at any location that they are working in. For Businesses, the Information technology departments have high demand to provide support for a wide array of applications on different enterprise software and mobility devices. They are also asked to modify existing relationships to meet the requirements of the industry that the enterprise operates in.

Businesses can use API management for a wide array of tasks. These range from maintenance of data to safely collecting and storing data which enhances the efficiency of the work and workers. APIs receive calls on them pertaining to different applications under use of the business and can provide other developers a chance to use your interface and your software to carry out their work and vice versa. Since APIs can provide the business with details regarding which applications receive the most calls, the business owners can find this as a new source of revenue through charging the developers for accessing and making use of their company's API.

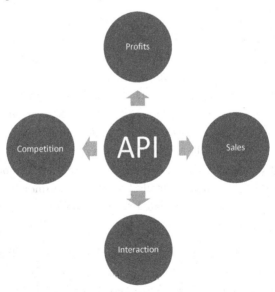

APIs can generate profits for the company that uses them. They help other developers and companies reach data assets of the company with the similar API and use the data to develop websites and applications. A company can let this happen while standing aside and watching, or it can take advantage of the situation and charge others a fee to be able to use their API and earn more revenue and profits from the use of this simple technology.

Other than providing the hosting company with a way to earn money through charging a simple fee, the use of APIs can enable a firm to achieve constant sales revenues. The use of APIs can raise the traffic on the business website and raise sales of the goods being offered online. For this to be achieved, the company should ideally not charge for API access to its website, as the information giveaway earns the company higher sales revenue.

The competitive structure of the business sector enhances the already immense need for adapting to API projects. The constantly emerging new platforms and ways to interact with programs and software do not allow for any lags in a company's willingness to adapt to new technologies and ways. To stay in competition with other market leaders, a company must invest in APIs to ensure the capturing and holding of a good size of the market. As competition is fierce, offering of data in the form of APIs is being offered already by all leading firms, or the process of API implementation is at least underway. Companies must cooperate with API systems to stay ahead in the competitive sphere.

Another very important aspect of using APIs is that the implementation of it ensures that other parties will make use of it. This is important because it helps ensure that the enterprise is going through healthy interaction with stakeholders. This helps the firm stay with all possible opportunities in the market. To help mobility and ensure that data can be made available on mobile devices for use by employees, a business wanting to be progressive and profitable will invest in APIs and support their adoption. It is suggested that data be made available in the API, which should be followed by using best developing companies and individual developers to help cover profitable platforms.

API management is an integral part of enterprise operations. It helps a company achieve financial advantages.

- In managing API, a firm gets to know the API users and what tasks they request from the company. This is good, as the company can have access to data regarding what kinds of people access their data.
- IN managing APIs, the company must set up proper modes of payment for those who wish to use their API and ways to monitor and restrict their usage conditional on the payment they make.
- It is also recommended that for API management, a company should monitor the number of times a person uses an API (if it is free). There should be a cap on this to discourage too much traffic in the company's profile at a single time.

Many companies offer API management services to those businesses who do not feel they can achieve efficiency in the area. These companies charge for their services in API management and report generation, data monitoring and API usage policies as defined by the Those that hires them. The most common form of payment for API management services is through commission on the API usage and traffic to the site.

Prevention of Unauthorized Access to Corporate Data

As a company goes mobile, it experiences numerous advantages. However many risks and disadvantages can present themselves in the form of theft and spying on data, decision making and special secret business information. The protection of the company's important data is important so it is necessary to take all necessary precautions which would allow unauthorized access to private and sensitive data.

This risk is great because as the development of enterprise mobility devices and software increases, the amount of people who have knowledge of it also increases. These advancements, however, put businesses at a nearly-constant risk for theft, wrongful representation and misuse of company data.

There is a need in the corporate world to have access to information and its databases and databanks. Apart from this need of the employees, the enterprise CEOs aim to achieve a competitive advantage over all others in the market.

The downside of using equipment and smartphones and other mobility devices is that the frequency of important corporate data being shared online is subject to cyber threats such as hackers and spies gaining unlawful access to corporate data and causing inefficiency in the work of the employees who are using these devices. International laws against cyber-attacks and cyber laws are gaining importance and recognition due to the sheer volume of such attacks taking place all over the world.

To ensure a successful mobility enterprise system, a proper data protection policy and strategy must be formulated. The main task is to enhance productivity without hampering it in any way, reduce the data and information related risks in a cost effective manner that is at the same time, efficient in its work.

The main area of data theft and data loss is through misplacement or theft of mobile devices. Even in the presence of such risks, companies choose to invest in mobility devices. The importance of these devices calls for a device management structure for data protection. This can be done in three steps:

- The first step to be considered by companies who want to ensure security of their mobile devices against unpermitted access and mismanagement that commences as a result, is to use Mobile Device Management Applications. The applications are not limited to iOS users as common belief wrongly dictates and mobile devices of virtually any make can benefit from the use of MDM applications. As it is a recent development in security and device management, it has a long way to go before it is deemed perfect. However, it is the best form of protection application that can be installed within a device to quickly recover data from a lost, stolen or crashed mobile device and to track the location of it. This can help enterprises keep track of their devices and wipe important data off them in case of misplacement or theft. One important aspect of these applications is that they can work for 'anyone' Icons such as Tommy Hilfiger use these applications to protect their intellectual property and business assets stored in mobile devices such as smartphones, laptops and tablets.

- Data encryption for communication is an excellent way to establish control over who access data that is being stored, obtained, or sent through a mobile device. The purpose of encryption is to code messages and data in a way that only the sender and receiver can access it through a special encryption code that both parties have knowledge of. No one else has access to this data and its protection is guaranteed, as a third party cannot benefit from obtaining the information, as it will make no sense to them without decrypting the data. If data encryption is used, the fear of unprotected internet access points being used by employees to carry out business operations will not be a threat to the firm, as tasks will be carried out in a coded and protected manner.

- The enterprise wishing to ensure data protection should invest in enterprise sandboxes and the use of VPNs for transmission of information in the company. VPN is an abbreviation for Virtual Private Network, which in essence is a set of processes that enable a private official network to span over the internet, or other large networks. Different devices can share data through public networks while maintaining the interface and security policy of the private network. When used over the internet, a VPN connection enables workers to attain access to company networks and data while they are on the go. This ultimate security method enables unwanted access to corporate devices to be limited. Companies can share information and transactions through mobile devices or computers at the office without the worry of unauthorized access to their secret data. The highest-level communication between governments is carried out with VPNs, which serves as a good representation of the efficiency of VPNs to secure information transfers among members of the firm.

To put these plans into action is a daunting task for enterprise mobility management, as the presence and working of an IT department in the corporation is not enough. The users (both the organization managers and employees) should know the policies of use of the device which is being provided by the company to them. Not reading the terms and conditions of use of private information in a device for application can result in increased vulnerability of the device, and put the company's data at risk from invasion and wrongful use. For the fair and efficient use of mobile devices, an enterprise should make sure to engage the workers in interactive training sessions and informative sessions to make them aware of the threats the outside world possesses for important corporate data. The company should also adopt a security policy to ensure workers pertain to the requirements of the firm in order to maintain protection of corporate data against unauthorized access and unlawful use.

Managing Security Policies

Every business that has its eyes on attaining mobility requires a security policy. Certain standards must be set and procedures should be defined in order to ensure the security of information. To come upon a security policy for a specific business, special policies must be considered and evaluated to identify the advantages, compatibility and tradeoffs of the implementation of the policies. Risks in the business world are not limited to just the physical boundaries of the workplace and must be reduced in all sectors of operation: at the firm and outside using public and private networks. Security at the corporate level is extremely 'important' as it ensures the protection of key data of the firm, its operations and work procedures.

Security Policy: Importance

To ensure good security policy management, three steps must be followed:

- The most applicable and efficient security policy should be sought out and implemented in a way to best suit the nature of the company and its operations.
- Before implementation of a security policy, the reactions of the workforce to these policies must be assessed to ensure there is no resistance from the employees.
- Finally, some services to manage security policy, implement, and direct changes and amendments in it

should be outsourced to specialists who deal with such activities.

The benefits of a sound security policy are vast. A benchmark is placed on the activities of the enterprise and its employees to ensure the company's information security. This raises the need for good work ethics and high standard of work, general security and compliance with regulations of the company.

In designing a security policy, the most efficient way of formulating and managing a security policy is to let the higher levels of management create the policy according to their perception regarding the company's security needs. The policy should be drawn up out of thin air, but should base itself on the standards of the industry in which the enterprise operates. The work need not be done by the senior managers themselves- it can easily be outsourced to specialist firms or use of specialized applications can be made to ensure the utmost accuracy and efficient design of the security policy.

Creating a properly defined and comprehensive policy is essential to a firm. While in the development phase, here are a few points senior managers must consider while sorting out policy details and requirements:

- Understand enterprise culture and needs. The managers formulating the security policy must make themselves aware of the structure of the enterprise and what aims it has for growth, work and sustainability in operations. The managers need to formulate a policy that is consistent with the relationships between stages of the organizational structure, the roles that are to be carried out at different departments and the understanding of the lower segment of the organization.
- The nature of the policy should be clear and identifiable. The aim of the policy (be it for intellectual property

safety or IT security) should be clear and known to all the members of the team who take on the task of formulating the policy. The managers should also know what the policy is not meant to cover; decided through a consensus among the team members.

- The policy should not be too simplified, or too complicated. The audience for whom the policy is intended for should be able to understand and implement the policy to their work routines and at the same time should not feel confused about the technical jargon being used in the policy.

- The policy details should be easy to implement in work practices. The policy should not ask for behavior that is too complicated to implement in the workplace. For managing this aspect of policy development, it is suggested that a focus group be identified and the policy be tested on their work in their specific department to see if it disrupts existing work practices or if it is consistent with them.

In changing thing in the business world is technology. Companies must stay current with all IT advances and possible threats and corresponding changes will have to be made in the security policy's IT section on a regular basis. This is a requirement of security policies that all firms must remain prepped for. Extremely sophisticated and efficient IT security policies are bound to become outdated within a certain period of time, and so to protect the company information and infrastructure against all possible threats, security policy requires firms to create a system to regularly analyze and update security policies of the firm. The IT aspect of a company's security policy, one thing must be considered. The most rapidly

A good security policy may in no way remain static in nature. Having dynamic policies open to change is a requirement of managing a good security policy in a firm. This need not be carried out inside the enterprise itself. This work too can be outsourced and efficient organizations can be given the responsibility of maintaining a dynamic and up to date security policy.

A security policy also asks for easy assessment after implementation. Several tools are sold in the market to measure the compliance and effectiveness of a policy and a policy must be designed and fully implemented to be easily assessed for compliance and success.

Application Management Requirements

Applications management refers to the dynamic ways to enhance, control and change applications in accordance with the dynamic needs of business activity. Applications management is required to enhance applications in ways to ensure that software do not fail to fulfill business needs. It involves the services for applications and the help for application systems during their life. Support may be offered to users of the applications and software used to run applications is bettered and perfected through applications management. It is basically a form of outsourcing technological services.

The need for Applications Management

Application management is required due to the rising complexity that application interfaces go through due to the high pressure they face from changing business needs coupled with limits on funding for development. Other than this, enterprises face constant change in practices, policies and cultures over very short spans of time, and applications need to be altered in accordance with changing needs and before being sent out into the business market, should be well tested for good performance and should be applied to multiple user interfaces.

The applications market itself is a highly competitive one where newer applications take over old traditional ones on a regular basis. Hence, applications management is necessary to make small and big alterations to make applications more competitive and applicable to growing interfaces and technological infrastructures. Lastly, the need for applications management arises due to the fact that providers of software consistently come up with new devices on which applications are to be used, so they need to be changed in accordance with the changes in the devices.

Requirements of applications management come in the form of the many challenges it faces in the corporate world. They require higher quality, an unmatched worldwide supply of services regardless of where operations take place, innovative technologies and ways to simplify business dealings, lower costs and increase profits from the application.

Providing the Necessary Middleware to Fulfill Requirements

Middleware is a term that is used to identify the kinds of software which are in essence used to enable data management and transfers across different platforms through the use of applications. It serves the purpose of a middleman in bringing two parties together without the need of a human to actually carry out the job. The work can simply be outsourced to the middleware that will act as an intermediary to fulfill the needs of applications and data transfers. Many services can be classified as middleware and include enterprise service bus, message oriented middleware and data integration etc. Middleware involves specialist structures that are strategically placed between applications and the operating system. It aids the building of applications for firms. Middleware serves as a translation software and integrator between two parties.

Computing system middleware can be used to provide services that take a lag to complete and make it seem like a human is being directed to carry out the task, or to perform a service in real time. Real time middleware operations are standard in nature and can be used across industries and commercial sectors.

There are two kinds of middleware: and.

- Middleware enables the transfer of data and corporate information among applications. It is quite often employed in customer relations applications used by enterprises.
- Middleware facilitates the authenticity of tasks carried out by the firm through secure transfers of corporate data and other information.

The two kinds of middleware are dependent on a server, where the information about the company is sent through several layers such as the customer relation layer, to the layer of the middleware where they are transformed into any interface the user in an application layer that the user can identify with and infer information from.

There are several other alternatives that can be chosen by companies if they do not wish to invest in middleware where programmers are requested to integrate different applications that can then be allowed to share information, but the data sharing is restricted between the two, and cannot span across to another recipient application. The problem this creates for enterprises is that of problems with flexibility pertaining to different business challenges that they are faced with. In the business world, the challenges businesses are presented with are great and need to be catered to with the utmost speed and accuracy. Absence of middleware does not allow for this; resulting in unnecessary lags.

The purpose of creating middleware was to provide assistance to different application infrastructures. Many enterprises have to work through several application infrastructures and interfaces. If this is the case for an enterprise, support should be provided through middleware to integrate different operating systems. It facilitates easy database transfers from one operating system to another.

As it provides integration among a company's devices, it aids the overall connectivity in the organization as it enables communication across platforms. Middleware enables API access to several databases in a simplified manner which is constant across all kinds of databases.

Middleware and its use helps companies operate without any lags in work virtually no slowdowns for employees in completing their tasks. Middleware troubleshoots all problems in varying application infrastructure and enhances the efficiency of a firm's operations.

Middleware helps customer relationship management in many ways. Since information runs a company, the most important data a company can possess is that of its customers. For companies to whom customers are of great importance (as is the case with virtually all business organizations), such data is what controls all company operations. The main problem company's face is transporting data from one source to another within the Company in real time. For this, reliable methods are required to transmit data without lags, hassles or discrepancies arising. A secure method to do such tasks is through the use of middleware.

This is a method that simply automates the CRM aspect of a firm and painstaking procedures are not required by the organization to convert data into compatible formats for different operating interfaces. The lag is greatly reduced greatly reduced and the cost to the firm of implementing a complete change in the company's systems to enforce data integration is minimized as all data can easily be used and transferred through middleware. Middleware, quite contrary to popular conception is very old in the corporate world. This is because the need to transmit large amounts of important data quickly and with ensured safety has been a need of all firms since the late 20th century.

The efficiency of middleware over the years has increased greatly, as the changes in technology from mainframe computers to now handheld substitutes for entrepreneurial use has provided developers and service providers with new forms of technology and better user interfaces to operate more efficiently and quickly.

Middleware is provided to those who need it through a web of countless companies operating all over the world, each of which specializes in one form of data transformation or the other. The oldest companies are the big players in the market for middleware and include companies such as Sun, Oracle, Microsoft and Motorola. Such large firms offer middleware in integration with the mobility aspect of business work, and aid many objectives of an enterprise with one software "middleman".

Application-server platforms is the key selling point of such companies as they offer ease of transmission and decoding and encoding of data from A business to the leading business software and applications available. This enables them to export and transport data to and from sources that are available in the application itself. What many large middleware companies do is concentrate solely on middleware, which provides them with great efficiency in the matter and helps them become specialists in certain fields and services that companies require from them. The success of the middleware industry depicts its importance in the business world, as it has reached multibillion US dollar worth in only a matter of decades.

The cost of middleware is directly dependent on the internal specifications of the enterprise. The factors that influence the price being charged for middleware include the traffic, the level or privatization and custom requirements of the client and numerous other factors. It is not possible to give estimates of the cost of middleware, as many factors affect the amount of work to be put into their development and thence the amount of money required for the process.

For customer relationship management integrated businesses, it is essential to have middleware to aid transactions and business processes. The integration of the systems with middleware should not be void of the aims and objectives of the client company's objectives. If the corporate culture of the client firm is not compromised, then middleware is a great investment and is likely to pay off its costs soon after implementation. If customer relationship management software is already installed in the systems of the business, implemented middleware is not a costly or time consuming feat. But, in the absence of this factor, the cost for implementation and use of middleware is likely to be high for an enterprise. Extra time will have to be put in and specializations will have to be made in middleware development to be able to meet the aims and objectives of the client requesting middleware.

Self-management Portals

A self-management portal is known as an Enterprise Information Portal, which serves to amalgamate data, employees and work procedures across the boundaries of the organization. Such portals are of extreme importance, as they provide company stakeholders with access points to integrate the different aspects of an organization. They ensure constantly updated and consistent data and information for those who use the portal be it the employees, consumers or the suppliers. They serve all forward and backward linkages of the organization, irrespective of whether they are a part of the organizational workforce or not.

Enterprise portals integrate and provide information and data to those who request it and can also provide other services such as integration of data and the ability to navigate between different components. Portals are usually easy to customize to suit the needs of the stakeholders in accordance with data requirements, company patterns or simply with the personality of the user.

The significance of these portals is in the fact that they can provide the enterprise with interaction with all stakeholders of the firm with a very important function with -- the ability to limit access to data. This enables controlled or limited access of information to employees or suppliers—giving no one the ability to gain unfair advantages to a company's valuable data.

Expense Management

In enterprise mobility management, it is estimated that over a quarter of an average company's budget is allocated towards the implementation of a mobility platform. This naturally results in an expectation of worth-while returns on the investment. Large organizations all over the world invest heavily in different device provider firms and brands to ensure the needs of their mobility plans are met.

The factors to be considered when devising a budget to control expenses for enterprise mobility management are as follows:

- What are the costs of acquiring and running mobility devices in the enterprise? The company must know all costs it will incur while mobility devices are used by the workers and should put an effort into gaining knowledge regarding the costs of carrier services across different providers. The choice for a carrier should be made according to one condition: the mobility needs should be met with the highest efficiency and at the lowest cost.

- Will workers need extra training to be able to use the mobility devices efficiently? This is a factor to be considered by the enterprise as it can cause great costs to be incurred if workers do not have any knowledge regarding how to use a mobility device.

- What is the charge for managing mobility from specialist organizations? This needs to be known to the firm to make relevant amends etc. to budget allocations and cost structures.

- To whom will the devices be given? There should be considerable thought before deciding which branch of the company, or which departments attain mobility devices. They should essentially be deployed to workers who are expected to move around a lot in their work, to ensure their high productive efficiency while on the go.

- Will there be wastage of devices? The devices to be chosen should not be weak or easily corruptible. The company should know what costs it might have to incur in making repairs, facing damage or finding the devices obsolete a few months after deployment into the organization. Such costs must be considered. The costs for recovery of data in the event of misplacement or theft etc. are also important and should be considered by the organization.

- The expenses that will have to be incurred in replacing older devices with new one must be accounted for. The best price should be attained for sale of old devices.

- The enterprise must ensure that it is prepared for additional funds to be placed to aid the troubleshooting processes should crises arise with compatibility and interface problems between the enterprise software and device interface.

- The company should encourage a BYOD culture in the workplace if cost management is an aim. This will require little or no training efforts by the enterprise, and the familiarity of the workforce with the mobility devices will be ensured. Here, there will be no tradeoff between the company's investment and worker productivity, and the company will only need to invest in a sound IT department to ensure cost management.

In accounting for these factors, managers can ensure the highest return on investment in enterprise mobility devices and manage the costs being incurred by the firm in an efficient and cost effective manner.

Planning for the Overall Cost of Mobility and Data Transmission

The most important factor to be considered for controlling costs in mobility management for an enterprise is that of data plans. One must be connected to a network at all times to:

- Constantly check updates and corporate email when not present at the workplace
- Enable conferencing with stakeholders of the business who might be at far off geographical locations
- Make use of cloud computing to transfer important data to clients or supervisors
- Any other enterprise needs

If such activities are required to be performed on a mobile device, permanent and constant connection to internet networks is required. The security threats of connecting to unprotected Wi-Fi hotspots in unsafe locations can pose serious harm to a firm, so most companies consider getting data plans for employees and their usage of Smartphone devices for mobile working.

Data plans are associated with a price tag that differs with the bandwidth that is acquired on a monthly basis. Other than this, special deals associated with the purchase of smartphones help enterprises decide a package for data plans for the smartphones they are purchasing for their employees. In such cases, the cost of data plan purchases is not very high. On average, a data plan with a 2GB limit is bound to cost the enterprise around the equivalent or current $ US 30. If the use of the employee exceeds this limit, a greater cost will be incurred for the data plan which the enterprise will have to pay for.

Depending on the usage of data plans on a device, the specification requirements vary with the amount of data required. The enterprise must consider service providers, Smartphone companies and the type of mobile data (such as 3G or 4G) that they wish to purchase for their employees. This can amount to a great cost for the enterprise, and a control on data plan usage needs to be exercised.

The company can specify the need for a limited data plan (this data plan stops users from using a data packet beyond the limit that has been agreed upon) to save costs. Moreover, the company can place a check on what activities a user indulges in while using the device. They can block certain data consuming applications and software from the phone to ensure maximum relevance of the data packet used. The e company can also install apps on the user's smartphone to measure and notify them of their data packet usage and limits (such applications are available on Windows, iOS, Android and are easy to implement all over the organization, even for BYOD policies).

SECTION 4 –

Other Aspects of Enterprise Mobility

CHAPTER 9

Mobile Device Management

What is Mobile Device Management Software?

As discussed in earlier chapters Mobile Device Management is about managing and controlling the information being downloaded, shared and utilized on the devices in use by mobile employees. These devices can be personal or given by the company. However, the main contributor towards a growing demand for MDM is the BYOD (Bring your own device) trend.

Initially the focus of MDM was to exert maximum control on hardware of mobile devices. Software's were developed to prevent misuse of cameras and Bluetooth. However, as the technology has evolved, MDM vendors have realized many of the potential security threats come from the apps on the device. There has been a shift of focus towards managing apps alongside the hardware.

By 2017 more than fifty percent of the workforce is expected to shift to mobile devices for remotely performing office tasks. While BYOD has received applause for reducing costs in the short term, in the long term the impact can be reversed if adequate MDM software is not in place. In such events, companies need to go an extra mile to make sure that employees remain mobile and corporate information does not end up in the wrong hands.

Given the rising demand for MDM, the IT industry experienced a stark rise in the number of MDM software vendors. In an ideal scenario, it is the end customer who benefits whenever markets become more competitive, however, in this particular case due to little differentiation between products customers end up confused.

Following the focus on secure means of accessing and dispersing information, vendors have come up with multiple MDM software with varying features and price tags. The question then arises, which MDM software is best for a given company? Well, there are certain factors that can assist in coming up with an optimal purchase decision. These include features like the configuration policy, remote control of devices and information, backup program and security shields.

A Solution for Managing Devices in the Workplace (MDM)

BYOD brings along countless security concerns that MDM software is expected to deal with. Before MDM it was the task of the IT staff to ensure that the mobile devices are used in best corporate interests. However, back then the mobile device usage was minimal. Since the percentage of mobile device users has increased, the need for large scale MDM has also increased.

The MDM software provides the company with the ability of managing, monitoring and controlling user activities on the user device, as employee work on the go. Evaluating how well MDM software solves mobility issues, asks for considering the control it allows, the security it confirms, the system support it provides and the reliable access it gives.

Evaluating control: one word vs. one word approach

The degree of control exercised by MDM software is determined by the approach towards managing information. It can be either a heavyweight or lightweight approach.

- *Approach:* This approach draws a fine line between the corporate and personal data and manages the former accordingly. The degree of control exercised is greater; hence, the MDM is more secure. It comes with a new interface and users have to be familiarized with it. It provides stronger prevention against data loss alongside encryption and selective data wipe capability. The approach uses an app containerization approach. All the corporate apps on the device are placed in a container and used as per the policy drafted by the company. Any violation of company policies can jeopardize all the apps in the container.

- One word *approach*: This approach manages the entire device usage (no differentiation of personal or corporate data) but is less stringent when it comes to controlling. It keeps the device's innate settings so users need not to be trained for optimal usage. A small control agent is installed on the device and most of the control comes from some customization of the settings already present on the device.

Evaluating Security

The very reason for implementing MDM software is to ensure secure mobile device usage by employees. Information shared on the company network can be secured by multiple ways including ability to encrypt data, selectively wipe data and blocking access to unauthorized users. An ideal MDM solution gives maximum control to the IT staff, allowing them to efficiently detect and remotely block any attempts of data attack.

The problem with the IT industry is that as the software vendors become technologically advanced, so do the cyber criminals. The truth that every software is susceptible to all sorts of cyber-crime, i.e. viruses, malware and hacking. The cyber-crime industry earns millions of dollars by stealing and disseminating corporate data. According to 'The Security Threat Report 2012' by Sophos, the Android market was flooded with hundreds of malicious applications, most of which were downloaded by thousands of users before Google could remove them.

The BYOD trend exposes the company to cyber-crime at multiple fronts. The market reports suggest there is no looking back in this trend. The coming years will see more of mobile device users in the workplace. Therefore, the only option left is to put in place stringent MDM software, which offers several security shields, automatic data wipe following a suspicious activity and instant notification of any attempts of breaking into the system.

Evaluating System Support

Only a few years ago, Blackberry was thought of as the only secure mobile device. It ruled the corporate mobility evolution with no competition in sight. The support system provided was effective and reliable. Then came the iPhone and the Android, which within just a few years, gained massive market shares. Company IT departments were overburdened with the task of managing multiple operating systems to ensure MDM. Common problems included slow system and delayed upgrades.

MDM vendors now provide solutions that can support multiple OS platforms. MDM solutions that supported a single OS have almost disappeared. MDM software hosting multiple operating systems provides a relief to the IT staff and makes system integration run smoothly. It is only futuristic to look for such solutions because the IT landscape is forever changing, who knows that a new OS may surface in the coming years.

Evaluating Reliable Access

Reliable access is a prerequisite of enterprise mobility. Thanks to globalization, companies have employees spread across the globe working within different time zones. The MDM software must be running round the clock at optimal speed. The MDM software exhibiting slowdown because of overload can lead to user frustration and reduced productivity.

Implementation of MDM

The implementation of MDM is a systematic process and is determined by the type of software being deployed. The requirements vary from investment in hardware to investment in skill development of employees.

However irrespective of the software type, the implementation phase is always preceded by a *need analysis*. This involves asking questions about the end user, the technology, the information, the cost involved 'the' with 'and' time considerations. For example:

- How many of the company employees are mobile device users?
- How many segments of users are there? How many interfaces need to be designed?
- How tech-savvy is the user? What is the technology adoption rate of the user?
- Who will manage the MDM if deployed? Is the support staff available or needs to be hired?
- How will be the access be given? Individual account or a single pin code?

- What information will be available to what user segment?
- Who is authorized to receive confidential corporate information?
- Which activities will be monitored on the mobile devices?
- Which operating system(s) are in use?
- What is the required capacity if an MDM is deployed? How scalable should it be?
- How much will the software cost? How much will the entire implementation cost?
- How much time will it take to have the software running?
- These are only a few examples of the questions that can be asked to gain an insight into the need for an MDM. The key is to move from general to specific questions. Once the answers are known, it is time to move to market research for potential vendors of the ideal software that fits well with the company requirements.

MDM Deployment Models

MDM solutions are offered in two deployment models. It can be either on premise offering or cloud computing.

1. On-premise MDM solution

An on premise MDM solution, as the name suggests, provides on-site service only. It requires infrastructure investment, which is often heavy on the pocket. The buyer needs to deploy the software on their own servers. This means the buyer also bears the cost of maintenance and regular upgrades. In a nutshell, if you have an MDM software deployed on premise then you will face cost as well as ongoing costs. The upside of on premise MDM solutions is that they come with maximum security and control over the mobile devices. Because of the fact that the

MDM software runs on a company's IT infrastructure headed by the company's staff, on premise deployment gives the user the privilege of enhanced management capabilities, whether it be for device or data.

2. Cloud Computing MDM solution or SaaS

The other option is getting an MDM cloud. For MDM cloud software, or SaaS as it is called, the good part is that the buyer bears no costs related to IT infrastructure investments. All the processing, storing, controlling and access management is done through the vendor's servers. Another factor is the system speed. The fast processing makes user access easier and management function more efficient. Moreover, cloud computing MDM software is not contained within the physical boundaries of the company. It offers greater access to users even in remote locations. Also referred to as 'on the cloud doc', this deployment model means less data loss in case of theft of the device. This is because the data resides in the cloud and the device is merely a connecting platform not a storage unit. As soon as the device is disconnected from the cloud, there are no traces of corporate data on it.

However, like all technologies, cloud computing has some potential drawbacks. Firstly, the degree of security is relatively lower as compared to on premise MDM software because of the simple reason that all information flows through someone else's server. A mitigation strategy for this can be the use of a private cloud instead of a public cloud. Despite the attempts by cloud computing vendors to ensure user security by enacting elaborate privacy policies, the concern yet remains. Secondly, cloud computing can lead to over-reliance on third party infrastructure for routine operations, therefore exposing the company to potential operational risk.

An Integrated Approach towards MDM Deployment

It was as recently as 2010 that market reports showed that the majority of the MDM users followed an on premise deployment model. Since then the trend has shifted towards a hybrid approach. Software vendors have been trying to integrate the popular cloud computing technology with the on premise MDM software.

Cloud computing was regarded as a disruptive technology in the IT industry at its inception. But does it perform the ideal function of MDM software? That depends on the type of company and the information being managed. An on premise MDM solution is a better choice for companies belonging to industries where the flow of sensitive information between employee and business is a routine. An example can be that of pharmaceuticals and financial firms. On the other hand, cloud computing MDM model is a good starting point for firms that lack substantial capital.

The focal point here is that both technologies have their own pros and cons, but if approached in an integrated manner they provide enhanced mobility management with greater security and access. Integration is essential in cases where there is a need to link the information on the cloud with the information in corporate data centers to increase validity and cut redundancy.

Implementing an On-Premise MDM Software

The implementation of on premise MDM solution asks for investment in IT infrastructure and IT staff. This is a time consuming process and can take weeks or even months. Figure 1 lists the main steps involved in the process.

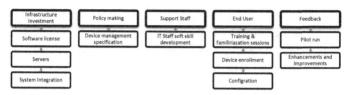

Figure 1: Implementation of on premise MDM software solution

It starts with obtaining the licensed software and running it on company servers (existing or new). The license policy can be *one user, one license* or *one device, one license.* The former allows a user to install the software on multiple personal devices. This type of product is preferred when the majority of the employees own and work on multiple personal mobile devices. The software is then integrated with the company's IT workflows and processes.

The CIO and senior IT staff members determine the terms and conditions of use. The interface is designed for multiple users at different organizational levels. The details on device management specifications are mentioned in the next chapter.

Then comes the training of the IT staff responsible for the maintenance and upgrading of the software. Companies with huge IT departments and experienced IT staff might not need external hiring. They can benefit from their existing human capital.

Apart from the maintenance staff the end user also needs to be educated, especially if the company is pursuing a heavyweight approach. This requires training and familiarization sessions. Deployment invitations are sent to employee devices. After entering the personal information the user device is enrolled into the MDM software with minimal role of IT staff. The device enrollment process authenticates the user and applies the required configuration settings into the device.

It is only wise to start with a pilot study before going full scale. It can be focused on one department or a few people from multiple departments. The advantage of running a pilot experiment is to gauge the suitability of the software and check its acceptance rate. It can highlight possible loopholes in the system. The findings of the pilot study are then incorporated while making enhancements and improvements in the user interface. Once done, the MDM software is ready to be installed company wide.

Implementing Cloud Computing MDM Software

As mentioned earlier, due to no infrastructure requirements the cloud computing deployment model is timesaving and cost effective. It can be on a public cloud or a private cloud. Experts suggest the choice of cloud to be based on the type of information that will be shared. For highly integrated complex processes and strictly confidential information, choosing a private cloud for MDM deployment is a better option.

The steps involved are fewer than for on premise MDM deployment. The company can choose between two types of options. One is that the company can get the software license for MDM vendor and run it on public or private cloud after registering with the third party vendor. Second is to buy the software from an MDM vendor who offers cloud-computing services as well. The second option is less as it has only one supplier involved.

Whatever the choice of product, the registration step is followed by setting up company policies for MDM cloud usage. The device enrollment and configuration are done in the same manner as in the case of on premise deployment model. Since the maintenance is the responsibility of the vendor, the IT staff needs not to be given special trainings. However, this can differ in case of deployment on a company owned cloud. The feedback step is also followed approximately in the same way. The deployment is completed within a short span of time.

Device Management Specifications

The specifications for MDM software are different for different companies and even within the same company, they can vary across departments. Recall that in earlier chapters it was discussed that the main purpose of installing MDM software is to ensure security and accessibility for employee mobile devices. The MDM specifications are set by the IT staff/admin to help achieve this very purpose. As mentioned in the figure the specifications can be divided into various categories for the reader's ease of understanding.

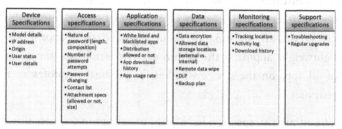

Figure 2: Device management specifications

Let's take a look at each category in detail:

Device specifications: The MDM gives access to details of the device from model specifics to a device IP address. These details serve the purpose of user/device authentication. They also help in identifying pin pack, stolen and jailbreak devices. Devices are given the user status of employee, guest or unknown.

Access specifications: User details allow the admin to segment users based on their department and management level. Based on these details, access is given to the relevant corporate information.

The very basic access specifications can be in the form of a pin code or a password. The admin can set the allowed number of attempts to enter password, password expiration duration, password length, password type (i.e. words, numbers, symbols), attachment size, log-in time out and the contact list, just to name a few.

Application specifications: Through a management agent, the MDM software configures the device with the set policies of the company. It automatically updates the admin of any attempts to alter the required settings.

The MDM is used to allow download of whitelisted apps and block blacklisted apps. The MDM software generates alerts if there are any repeated efforts to install a blacklisted app. This is a measure to prevent malware and encourage employees to download apps from a secure organizational source only. It also gives the admin information about the usage rate of apps. In the lightweight approach the admin exerts control over the usage of all apps on the device whether for personal or professional purposes.

The admin can specify a list of apps to be kept in app containers and downloadable apps from the company app library in addition to checking app download history, viewing app usage and stating compatible OS.

Data specifications: The data management asks for specifications regarding data encryption, downloadable data, copy paste restrictions, allowed data storage locations (I.e. External storage or internal storage), remote data wipe, data loss prevention (DLP) measures and backup plan in the event of data attack.

Monitoring specifications: The admin can monitor the device location, activity logs and download history for any mobile device user. The acceptable range is set for each user group. A breach in the set specifications can show a red flag to the admin and initiate prompt action.

Support specifications: The optimal functioning of MDM demands a consistent and reliable support mechanism so users can inquire about any issues they face. The specifications include the user guides for troubleshooting, availability of support staff and contact information of support staff. For on premise MDM deployment, the support staff is the company IT department, whereas in the case of cloud MDM deployment the support is provided by third party vendors.

These are general categories of device specification management; within these subcategories also exist to exercise greater control.

The Uses of MDM in the Enterprise

It is said that the only thing constant in life is change. This goes well for the IT landscape. Consider the MDM software which received attention for making mobile devices secure but as time passed, this function is more like a given and the buyer often ask *what else?* Customers want more for less and higher value for the same price.

Now coming back to MDM software, by now it has been stated and elaborated how MDM can serve the company's interests for secure mobility. Typically, MDM software is expected to ensure secure information flows to and from employee mobile devices, but in addition to this basic use, MDM can perform some other functions as well. Vendors keep on innovating to enhance their product's value proposition. Innovation does not necessarily mean incorporating a breakthrough technology. It can be as simple as greater convenience for users or some kind of impressive analytic tool.

Telecom Expense Management

Also referred to as TEM, telecom expense management tools are provided by vendors coupled with MDM software. TEM was previously thought as a separate function but now the scenario has changed. TEM along with MDM can provide enhanced mobility security for a company. While the TEM can offer insight about the call logs of the users, the MDM can provide an activity log on the device together the two management tools provide control over all potential mediums of data attacks or data dissemination. However, because of privacy concerns employees may show reluctance towards the company's decision to adopt such an approach. After all, who would be comfortable with being monitored all the time? However, despite these concerns, the practice is widely appreciated for providing extensive control.

Coupled Services

Many antivirus vendors, looking at the demand for MDM software, have hopped into the scene and started offering MDM solutions with added security. Since the expertise of antivirus vendors lies in device protection, they capitalize on their existing capabilities to enter the growing market of MDM software. An example can be that of McAfee Enterprise Mobility Management (McAfee EMM). McAfee is a well-known name in antivirus market and it was only as recently as 2011 that it introduced its product for MDM purposes.

A company that has McAfee antivirus installed in its system will face little or no trouble while implementing the same company's mobility security software. This adds a convenience factor for the buyer.

Mobile Application Management

Also referred to as MAM, the mobile application management is deemed to be the new front for competition amongst MDM vendors. Market insights suggest that third party apps are used by as much as Fifty-one percent of smartphones owners. So despite the fact that the MDM software can manage the mobile devices it does not manage the apps on it. Hence MDM alone cannot provide sufficient control over enterprise mobility 'Another' level, i.e. mobile applications also need to be monitored.

MDM software vendors in collaboration with independent software providers come up with apps that meet the safety requirements. These apps are then offered alongside the MDM solution for better control. Many MDM software vendors have now started offering MAM to provide fool proof mobility security. 'Any ware' by Mobile Iron and 'Air Watch Workspace' by Air Watch are examples of the growing trend towards MAM.

These are just a few examples of the prevailing uses of MDM in the current IT environment. MDM vendors should learn from history and keep the value-added services coming if they intend to retain their existing customers.

Technology Insight: App Wrapping to Outgrow App Containerization

While the admin sets the specifications for app usage, it is essential to make a choice between app containerization and app wrapping.

App containerization is a relatively old concept widely used in ensuring MDM. App containerization is done so that apps on the device are used in the defined manner as per company policy. As the name indicates, it puts all corporate apps on the device in a 'container', which is accessible only through company stated terms and under the stated usage policy. All other apps on the device are exempt from the control. The major drawback of app containerization is lost native experience for the user, as the apps in the container can no longer be used. Moreover, containerization cannot be applied to apps provided by vendors leaving a hole in the security net.

This hole is filled with the new trend towards app wrapping. App wrapping is a product of the trend towards Mobile Application Management (MAM). It asks for bringing security control down to app level instead of device level (as in the case of app containerization). Since the MDM vendor itself provides the app, the mobile devices experience superior security. App wrapping does not disrupt the user experience.

Market analysts anticipate faster growth for app wrapping in the coming years with a growth rate of Twenty-seven percent until the year 2018. Who knows, by then a new trend will have surfaced in the mobile device management field.

Chapter 10

BYOD

The Concept of 'Bring Your Own Device'

To understand the concept of BYOD, it is important to look at how the industry is responding to it. Below is the success story of Intel Corporation's BYOD program.

Industry Insight: Intel's approach to BYOD

Wise people say the sign of a great company is its approach towards events. Great companies build leaders who have a futuristic approach and are not reluctant to be innovators. Intel a 105.72 billion USD[20] company fits this definition of a great company. Intel is part of the semiconductor industry and ranks 6th amongst the world's most powerful brands as of 2013.

BYOD at Intel

Intel officially introduced its BYOD program in 2012 with 'trust' as the main driving force. Intel's CIO Kim Stevenson is

Company Overview 2013	
Employees	105,000
CEO	Paul Otellini
Sales	$53.34 B

a firm believer in trusting employees with information and

---

[20]   As of May 2013

technology. 19,000 employees registered with the BYOD program to benefit from the flexibility it offers. The implementation of the program involved attending to the tiniest of matters including support systems, policy making and handling legal matters.

The company deployed a private cloud to manage BYOD. Users can access company information and other services like instant messaging through the cloud. To promote effective implementation of the plan, the following measures were adopted:

- Deploying electronic peripheral vending machines and help stations in the offices so employees could get support whenever required
- Designing awareness programs for employees. These include four session courses for new employees, whereas old employees receive yearly refresher courses.
- Deploying Mobile Device Management (MDM) software to ensure compliance and security of mobile devices
- Planning an employee exit policy Which includes data wipe and inactivation of certain applications whenever the employee leaves the organization
- Setting two Wi-Fi networks with -- one for employee use and the other for general use. A user of the latter network has no access to corporate information
- Giving access to information based on trust zones. The user profile, usage context and the device determine the trust zones. The zone determines the information content a person will be able to access.

These measures formulated a corporate culture promoting technology and freedom. Intel's CIO believes in trusting the employees instead of being paranoid about the risks that come along with BYOD.

The challenges

The major challenge was to carry out electronic discovery (e-discovery) on the user devices, especially in such an intensely regulated industry. E-discovery refers to the exchange of information in case of legal affairs. The information format in e-discovery is referred to as Electronically Stored Information (ESI). Multiple user devices, also referred to as Small Form Factor (SFF) devices, make it difficult to cater to the request for e-discovery and access ESI. The major setback is that since the company does not own the devices, it exerts little control over the information stored on them. Intel dealt with this challenge by implementing assertive policies and following industry best practices while laying out the blueprint for a BYOD plan. This approach enabled Intel to successfully pursue its BYOD plan.

Another aspect was the widespread concern over data security, but according to the CIO of Intel, all technologies, even PCs and tablets, have such risks, so it is not wise to abandon this trend from the fear of potential mishaps. She believes that putting 'trust' at the core of the BYOD plan and focusing on people development can help overcome this challenge.

The results

The outcome was encouraging. By the end of 2012, the plan covered 23,500 mobile devices at Intel. Speaking in financial terms, the CIO stated that Intel earned 3 USD per 1 USD of investment in this program. The per day productivity gain was 57 minutes, accumulating to annual savings of 150 million USD in the first year of implementation.

Following the success, Intel is now pursuing expansion in the BYOD plan. It plans to add more services and applications to the cloud. Work is already being done for offering instant video conferencing on the cloud.

BYOD in the Business World

Having discussed the case of Intel let us now look at what market research says about BYOD. Market research claims that BYOD market will grow at 15.7% annually and amount to 181.39 billion USD in 2017. Back in 2011, the market was worth 67.21 billion USD. Since then the trend has received wide acceptance by users and companies across the business world. The largest share in this growth is held by the North American region where the BYOD market is expected to jump from 24.26 billion USD in 2011 to 58.6 billion USD in 2017.

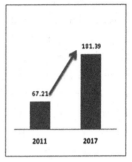

Figure 3

BYOD across economies

In 2012, a study performed across seventeen countries by Ovum, a research firm, found that the BYOD trend is followed differently in different geographical regions. This was mainly Due to differences in the growth rate of the respective economies. In emerging countries, the BYOD trend grew faster than in mature countries. Employees working in emerging countries were eager to have full-time access to the workplace irrespective of their location, whereas in mature economies employees appreciated a fine line between work and life.

This has contributed to a greater use of personal devices for office work in developing countries, hence a greater need for mobile device management (MDM) in these regions. The study also revealed that in regions where the BYOD trend had high penetration, the company IT departments tend to exhibit greater knowledge about managing mobility. Moreover, in countries where the trend is relatively new, the companies lack any policy regarding BYOD, which makes them a susceptible target for hackers seeking access to confidential corporate information.

BYOD and device/ OS choice

As a particular trend gains momentum and starts to penetrate the market, some informal rules start to surface. Same goes for BYOD, users start linking it with a certain type of mobile device or OS. Take for instance the choice of OS. Due to enhanced security features and lesser susceptibility to malware, Apple operating system is recommended by Fifty-eight percent of Businesses to their employees with -- according to a market report by Centric System. The second preferred choice in BYOD is Android.

The choice of device and OS by BYOD followers can also depend on the type of industry. Since every industry differs in product type, business processes and user experience, the communication requirements are different as well. For example, as per a report by Citric, Android OS dominates the communication services industry whereas iOS is the top choice in legal services and insurance industry. The reasons for dominance of a certain type of OS can be based on functional compatibility or user referral.

The Introduction of BYOD into the Market

The trend towards bringing personal devices to work started almost a decade ago. The introduction and continuous innovation in the smartphone industry fueled the desire to use a single phone for personal and corporate matters. Let us look at the events that led to the BYOD trend as we see it today.

Blackberry: The trendsetter

Although the smartphones is considered a product of recent technology, the term itself is quite old. Ericsson first used it back in 1997 when it launched the GS88 mobile series.

If one is to indicate the start of BYOD, then undoubtedly the title of trendsetter goes to the Blackberry, the brainchild of Research in Motion (RIM). RIM blackberry had been present in the mobile industry since 1999 but it was only in 2003 that it made its mark on enterprise mobility. The 2003 Blackberry model was part of the 6200 series. It had a monochrome screen with a QWERTY keyboard and functional features like email access, web browsing and greater memory.

The color screen version of the Blackberry, the 7200 series, was added the same year. For the first time, employees could access corporate emails on the go. All the top-notch executives had a Blackberry. It was the 'in' thing. By the beginning of 2004, Blackberry made it to a million subscribers. The competition was minimal and came mostly from a company named Palm. For years, Blackberry held the title of being the pioneer of enterprise mobility and being the top choice for business executives.

The rise of smartphones and mobile devices

The focused positioning of the Blackberry left room for competition to enter and cater to the requirements of non-corporate users. Competition hopped onto the scene and smartphones positioned as *not only for corporate* emerged. IT industry received a breakthrough innovation by Apple in 2007 in the form of the revolutionary iPhone. BY 2008, RIM's share of smartphone market was at Fifteen percent while that of Apple was at Thirteen percent.

The devices kept on getting smarter and smarter. Since 2007, customers have seen new smartphone products from Microsoft, Nokia, Samsung, Apple and Blackberry. Email and web browsing were no longer added features; instead they became a norm. Now the user experience drove competition. According to Gartner, the sales of smartphone exceeded the sale of PCs in 2010.

Another contributor to BYOD is the tablet PCs that hit the market in 2010. The Apple iPad, Google Nexus and Samsung Galaxy tablet all thrashed the sales of traditional desktop PCs. Mini laptops clearly lost the competition to these tablets.

Having stated the history of the rise of smartphones and mobile devices, it is essential to note that the added features these products had made them suitable for both personal and official use. As the user grew fond of a personal mobile device, they started giving it preference over company allotted mobile devices. Slowly the BYOD trend had its roots grow deeper into the business world.

By 2010 the industry analysts held different views about the BYOD trend. Some viewed it as nothing more than a fad, while others thought of it as a sustainable trend. However, time proved that the BYOD trend was here to stay. The recent years have seen a rise in BYOD penetration despite the associated security concerns.

The contemporary world

The world was going through a connectivity reform and everything in the IT industry was being redefined. The consumerization of IT was caused by growing internet access and greater penetration of smartphones. Let's take a look at some market statistics:

- According to Comscore, an internet research and analysis firm, the number of smartphones users in the

US rose from approximately 10 million in 2007 to 110 million in 2012.

- As of 2012, out of the total 5 billion cell phones around the globe 1.08 billion are smartphones.

- According to Neilson, in the US the number of smartphone users accessing mobile Internet increased 45% between 2010 to 2011.

- According to research by Enterasys dated 2012, the number of iPads sold in 2 years exceeded Macs sold in the last 20 years.

According to multiple research resources compiled by Track Via, Eighty percent. Of employees use their own devices for business purposes.

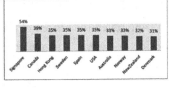

Figure 4

- The highest penetration rate of Smartphones is in Singapore at Fifty-four percent the figure 2 indicates the Smartphones penetration rates in multiple countries. Smartphones are kept in use round the clock by a striking Eighty-nine percent of owners.

Given the statistics mentioned above it is justified to say that the technology reforms in the last two decades were much more intense than any technological leaps in the past. As smartphone users grew, so did Internet penetration. Staying connected became more important than ever before. People were not going to let physical distance stop them from working remotely.

It is by all means safe to conclude that the technological advancement and the consumerization of IT contributed towards the trend of using one's own device while at work. The comfort and flexibility offered by working on a personal device is not paralleled by the experience offered by company owned devices.

Currently, it is right to say that the BYOD trend is in its growth phase and will show sustainable growth in the years to come. As users grow fonder of their own mobile devices, the chances are that the near future will witness a rise in the number of employees using only one device for both personal and corporate use.

Risks Involved in BYOD

As the BYOD trend flourished, IT departments faced increasing pressure to ensure safety of corporate data. Though BYOD has benefits, it has some inherent risks as well. Stated below are some common risks affiliated with BYOD.

Lack of knowledge about BYOD

A research by analyst firm Ovum[21] in 2012 showed Seventeen percent of the respondents claimed their company IT departments had no idea they used a personal mobile device to access office data. This signals danger. The implications of not knowing how many mobile device users exist in your company can be severe.

According to Forbes, less than Ten percent of companies hold substantial information about the devices that access their network. This means the other Ninety percent either stays in the dark or is not fully aware of the consequences of ignoring this trend.

Cost of compromised or lost devices

In the event the device is lost or compromised, both owner and company bear some costs. The costs can be extremely high if there is no management system in place; having tangible as well as intangible components. While the face value of the device sums

---

[21]   The research spread across approximately 3000 respondents in 17 countries.

up the tangible cost, the intangible cost is that of lost information. Quite often the latter bears serious implications.

The cost to the mobile device owner includes loss of personal data, including contact lists and personal media. The more serious aspect is the loss of passwords and bank account information. Research shows that 3 out of 10 smartphones users claim to have sensitive information on their Device that ideally should not be accessible by anyone else. Despite security risks, many users gain access through unauthorized networks, jeopardizing their own safety. A study by Motorola to gain insight about user smartphone behavior found that less than 6 out of 10 people would reset their work related passwords in case the device is lost. This shows a laid-back attitude that can expose corporate information.

The cost to the company is mainly intangible. The leaking of sensitive corporate data can do massive damage to the company, for example, if any information about a marketing campaign leaks to competitors. In such a scenario, the research effort, time and money invested all go down the drain.

Another cost incurred by the company is when the BYOD does not translate into a productivity rise, rather acts as a mere distraction. According to research by Aberdeen Group, the company may also bear a cost in the form of lost discounts that could have been earned while buying the IT systems (i.e. Servers, printers, PCs) in bulk for office use. Since higher BYOD means a lesser requirement for office PCs, the company may not be able to avail discounts on IT purchases.

The cost incurred strongly varies across companies and industries, but the point remains that if ill managed, BYOD can shoot up the company's costs.

Combating malware at multiple fronts

Allowing employees to bring their own devices means the company has to secure numerous devices instead of one centralized system. According to Gartner, it is estimated that by 2014 the cases of malware attacks on personal devices would be double that of enterprise allotted devices. Since employees save work and access office mail from personal devices and not company PCs, the company has to make sure that these devices remain protected from malicious apps.

Statistics show that personal devices are more vulnerable targets for malware attack due to lesser security. The reason is obvious: enterprise devices (i.e. Desktop PCs, tablets and other mobile devices given by the organization) live under the shelter of strong antivirus software and security measures by the IT department. The apps on these devices are downloaded from secure and authentic sources. On the other hand, personal devices include apps for personal use as well, i.e. games, which are often installed from unauthentic and suspicious sources. Research shows that only Twelve percent of smartphone users regard phone security a top priority and the rest do not give this issue due attention.

A 2011 study by Numerates Research found that numerous companies have a different security policy for laptops, mobile phones and tablets despite the fact that all fall under the category of mobile devices. The laptops were subject to protection against malware and virus attacks, whereas tablets and cell phones were mostly given theft protection. All these risks tend to materialize in the absence of a well-planned BYOD policy. As evident in the case of Intel's BYOD plan, it is essential to mitigate risks instead of avoiding the opportunity that BYOD offers. The next section discusses the need and means to manage BYOD.

Making BYOD an Efficient and Safe Option

BYOD can undoubtedly lead to substantial benefits only in the presence of a well-laid policy. A research by Cisco Internet Business Solution Group stated that a well-planned and comprehensive BYOD program could yield financial benefits for the company. The study revealed that a saving of 3159 USD per employee could be achieved given proper management of BYOD.

Need for managing BYOD at your company

The need for managing this trend stems from the fact that Eighty percent of employees are using personal devices for work related activities. This is not a small number that can be ignored. It implies that a significant part of your corporate information is stored on mobile devices and you would definitely not risk losing it to the data attackers.

Secondly, companies need to cater to employee preferences to get the maximum out of them. Research compiler Track Via states that Forty-seven percent of executives consider BYOD a contributor to employee productivity. Another survey performed by the enterprise mobility firm, pass, showed that employees who are permitted to work from their own devices give as much as 240 more hours to work on an annual basis compared to those who do not work on personal devices.

Thirdly, a properly managed and implemented BYOD program takes the maintenance burden off the IT department's shoulders allowing them to focus on core business. In BYOD, the user bears the cost of maintaining and upgrading the device.

Finally, every business in the industry will eventually implement BYOD, so why not invest in it right away. A 2012 survey study carried out Cisco Internet Business Solution Group of 600 IT firms revealed that as much as Ninety-five percent firms allowed

BYOD in some form. BYOD was termed is very positive to somewhat positive for company operations by Seventy-six percent respondents. These findings shed some light on the industry's future prospects for this trend. BYOD is here to stay.

Working for a safe and efficient BYOD adoption: Industry Examples

Having stated the need to put in place a BYOD management policy, let's now discuss some industry practices for making BYOD safe and efficient.

Name: New York Law School
Industry: Education

Following the rising trend of BYOD amongst students, the New York Law School invested in a network access control product by Fore Scout. The product named Counteract performs the following BYOD management functions:

- Recognizing and authenticating student devices the moment they pass through the Service Set Identifier (SSID)
- Blocking any mobile device found to have malicious application
- Blocking P2P (peer-2-peer) sharing of documents to manage documents with copyrights
- Assisting in solving technical issues related to the device

The benefit of deploying this system was the relief given to the school's IT department. The system bears the load of managing the activities of 3,700 mobile devices on campus.

Colgate Palmolive
Industry: Consumer goods

In 2011 realizing the need for higher connectivity for better business, a BYOD program was initiated at Colgate Palmolive. The aim was to provide flexibility and mobility to employees as they perform office tasks instead of gluing them to desktop PCs. Employees had their own devices but no BYOD policy was in place to manage it.

Colgate Palmolive utilized the software 'Traveler' provided by technology firm IBM. Employees were to register themselves through a website, which then granted access to company email and contacts.

Colgate Palmolive had 2,500 registered employee devices in the same year. Financial calculations showed that the company yearly savings would accumulate to 1 million USD since it no longer had to pay corporate license fee for 524 blackberry users who had now registered for the BYOD program. The program gave employees the much-needed freedom to work irrespective of physical location.

Kraft Foods
Industry: Food and Beverages

Kraft Foods implemented its BYOD plan in 2010. However, the approach differed from that of Colgate Palmolive and NYLS. Kraft foods chose to have a stipend based BYOD program, whereby the company gave a stipend to the employees to buy mobile devices. The employees were to choose from prescribed devices. The plan was extended to 800 employees, but it did not cover certain managerial levels (i.e. Executive and factory workers).

The effort by Kraft Food was more like an attempt to promote BYOD in the company's corporate culture and enhance employee mobility. The only task performed by the IT staff was to allot stipends to the employees. The Chief Technology Officer at Kraft Foods, Mike Cunningham, stated that the plan was not meant to cover the entire company rather it was focused to a small part of it.

Every industry has a different take on this matter, as shown in the examples stated above. However, the focal point is that companies are following and embracing this trend in multiple ways to seek one benefit, which is enterprise mobility. Productivity enhancement is a mere outcome of this attempt to mobilize the company.

Making a BYOD plan

The BYOD plan makes it easier to reap the benefits of this trend. Irrespective of the industry some common steps can be followed while drafting a BYOD plan.

Figure 5: Making a BYOD plan

Firstly, it is important to Record which type of devices that will be covered under the plan. This may require doing some internal research to see which mobile device is most commonly used amongst the employees. Though research suggests that Apple leads the BYOD market, it is wise to cover multiple devices for employee satisfaction. According to the study by Cisco Internet Business Solution Group, employees gave top priority to being given the right to choosing the device they plan to bring to work.

Secondly, usage policies need to be set keeping in mind the company's interests and employee reservations. This includes the right to block certain devices, manage application download, allowing limited access and wiping data whenever deemed necessary. It is very important to explicitly state the policies to the employees so they know what they are opting for. The policy should also graph user segments depending on the information need. Maximum transparency increases the trust between employee and company. Awareness sessions do a good job of conveying the policies to the employees.

Thirdly, a company must decide the most suitable form of technology platforms for implementing BYOD. It can be an access control network, as in the case of New York Law school, a company owned website like Colgate Palmolive or a private cloud as deployed by Intel Corporation. A cost benefit analysis can serve as a good guide while choosing which option to proceed with.

Fourthly, the support systems need to be in place. Employees must know where and how to seek guidance in case there is any issue. Although the BYOD plan reduces the role of IT staff but it is essential to have a company help desk for better employee experience. Finally, evaluating measures should be in place as the plan is implemented. Usually the top benefit of implementing a BYOD plan is to increase productivity but performance standards exist to measure how effective the plan has been. This can help evaluate effectiveness and make improvements.

Chapter 11

Mobile Security

With the ever-increasing level of mobile phones used in the enterprise for corporate use, the need for protecting them rises too. Mobile phones have become an enterprise necessity and are now an integral part of the working of any organization, which is why companies and workers must know that the need to protect their mobile phones is as important as protecting a personal computer.

The Importance of Mobile Security

Think about it. Today, in almost every company that is operating in the market, almost all employees own at least one phone each. The chances are that when an employee so much as walks into the firm's office with his mobile device, his mobile will automatically connect to the company's network. In the event that the mobile has malware or any other malicious software in it, the network may become subject to its impact, exposing it to countless threats.

The threats the company becomes subject to are vast, and can impact several areas of operations. Viruses and spyware can expose the company's data, its work procedures and important information about the firm to parties that the firm wishes to protect it from. The use of mobile phones has given rise to the need for addressing these risks, and the need to draw up proper security plans and adoption of security software. The company will also need to educate its employees regarding the need to protect their mobile devices and how this is to be done.

The use of mobile phones and threats associated with them creates the need for a strong wireless network inside the organization. Naturally, no company will have an open access network operating in its buildings as data and information security is important, but it will not do to merely have a numerical security pin known to all employees. Individual authentication is essential, so that certain devices can be allowed or denied access; depending on which employee the device belongs to.

The high risks also raise the need for a close eye to be kept on mobile devices. For a company, this may be a daunting task as there are contractual limitations on tracking and/or wiping devices that are suspected to be used for unlawful means.

One step that larger companies take to prevent misuse or theft of corporate data is to implement mobile device management (MDM) in the organization. Due to data storage limitations in mobile devices, most employees resort to keeping company information on a cloud. Using a cloud application from Apple or Google is common among private users but this may not be a good option for corporate use. Most companies face conflicts with the terms and conditions of use of cloud applications. Hence, the investment in MDM is made by the major corporate giants to prevent data from being stored anywhere outside the firm. However, smaller firms face a big problem, as MDM is very expensive to implement.

Because most people don't understand the complexities of the security needs to protect devices against cyber-crimes and identity theft, hackers and cyber criminals, ' they' take advantage of this fact to misuse devices and the information stored in them. It takes hours of manual effort to correct such mishaps in the mobile device environment, which resultantly harms the financial position, emotional stress levels and even the reputation of the user who has been affected. This has raised the need to improve mobile security levels and strong password protection of devices, to mainly secure company data and information that is intended for private access by the company's members only.

Mobile device security can be harmed through the downloading of mobile applications that contain malicious content, weak device passwords and insecure network connections. When applications that contain malicious code are downloaded and launched to a smartphone, the attacker can access secret corporate and personal information and use it for a negative purpose. The most affected makers of mobile devices are Google, as their Android market is flooded with malware applications that need to be avoided at all costs.

Other than this, security through passwords and tracking systems is necessary to protect mobile devices. If a device is stolen or misplaced, the pass code should be strong and impenetrable. Other than this, tracking and wiping the device should be an option available to the user, to firstly locate the lost device, and if need be, wipe the user's important information from it. This will ensure that no identity theft occurs and that important data is safe.

In mobile devices, security should be a fundamental concept to prevent the misuse of devices and data by cyber criminals and hackers around the world. The need for mobile devices in all sectors of society is growing rapidly, thus the use of such platforms should be made secure and reliable for use.

Manufacturers must make sure mobile devices and applications are and properly designed; keeping in view all the threats, they will have to battle. Users should take relevant steps to protect their devices further against cyber-attacks and the risks they pose to the security of the user's private information and life, and that of the workplace.

The coming years present greater and greater security challenges for mobile devices and their users. Cyber threats are increasing in complexity that can confound the best software specialists, let alone layman mobile device users. Threats include stealing important data such as identification details of an individual, intellectual property and corporate data. The cyber-crime occurring through government and personal spies is also a great threat to not only corporations, but also national security.

Phishing scams and malware attacks are likely to also see an increase in the coming years as advances in technology continue to make traditional methods of safety redundant. This is why governments and institutions now feel the need to conduct awareness programs to educate people about the safety of their mobile devices.

The Challenges of Mobile Security

Efforts to enhance mobile security in any system, but they do not always meet success. There are countless challenges when developing and implementing mobile security strategies. Many actors are at play in the and will be discussed in the following passages.

Workers

Workers, the agents who make use of mobile devices, may sometimes pose threats to the company's security by their actions and use of the device. On average, every two out of five workers who make use of mobile devices claim to use them for personal use and for business work. This is not a problem as long as proper security measures are implemented in the organization and on the devices being used. If worker cooperation is achieved, devices may not be a threat to the company.

However, a staggering 4/5 of an average workforce is expected to use their mobile devices to access company networks (VPN and Wi-Fi, for example) without the knowledge of their employer or supervisors. This unauthorized access is not expected to happen occasionally, but on a regular (maybe even daily) basis.

Some workers have been reported to confess they take the liberty to access corporate data that is not intended for their use. The reasons for doing this vary, but are mostly to advance the workers' own personal interests, for example, to appear up to date and proficient to the supervisors in hopes of getting rewards and promotions.

Regardless of the brand of a mobile phone, malware attacks are common among all, except maybe Apple's iPhone and iPad, which operate on a relatively more secure iOS platform. The sort of spyware that plague such devices have the ability to gain access to all the data in a mobile device, spam the phones on the user's contact list and tap conversations on the phone.

The reason for this is the vulnerability of the devices under use, and a lax attitude towards security. It is an open invitation for the attackers to come and take what they need, leaving the affected users and companies with massive financial and time losses. The best solution is said using applications for security in all devices. This has been necessitated due to the fact that mobile devices are continually connected to a multitude of different Wi-Fi networks and providers.

The Consequences of a Malware Attack

Malware started as seemingly harmless and inferior software with limited ability to harm computing devices. When this threat came about, it was dismissed and expected to die out soon. However, in a very short span of time, malware grew in sophistication, and is now as widespread as the number of access points available to a mobile phone at one time.

For Malicious malware to infect mobile devices is no longer difficult, as the sheer volume of mobile users in the world (especially the corporate world) opens the opportunity to select from an pool of possible victims (about 2 billion users all over the world). The truth is that there are many mobile phones with closed systems that are not subject to such attacks, but the number of people switching to smartphones with greater access to outside sources is ever increasing.

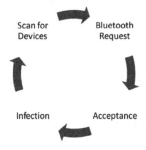

Using e-business, email, MMS communication and other online and network communications, users unknowingly expose their phones to malware attacks.

Bluetooth

Phones with Bluetooth access are susceptible to malware attacks as Bluetooth serves as a passage to transfer malware from one device to another. An unknowing mobile user who has malware in his mobile device can infect countless others merely by Bluetooth. His phone will search for other devices, spam them with request to accept a connection and upon acceptance, will infect the other devices too. The newly infected devices will now act similarly and spread the malware to other devices in the vicinity. The trick is to switch of Bluetooth discovery in the mobile devices to prevent this attack, which unfortunately most users do not make use of, thereby exposing them to malicious software attacks.

It is said that smartphones could well be the future computer for most users, and a large number of people who do not have experience with phone usage will be using them as substitutes for personal computers, and exposing their data to unknown threats and attacks.

Cyber Crime

Though malware on personal computers initially started out only for creating difficulties in operations, it is now a huge profit-making network for the criminals who develop it. Malware is used to steal important corporate data from corporate mobile devices, business trade secrets and even resources from computers.

The sources of cyber-crime are untraceable as they mainly make use of "Botnets", which are other hacked devices, to deploy the crime ware to other devices. They obtain company data and can sell it to make money, or threaten and blackmail the user (or company) to block their servers, shut down operations, etc.

This crime is not only limited to obtaining and misusing data of the corporate firm. Malware can also impact the costs of a firm by artificially inflating the bills incurred by users. For company devices, if malware is to impact some devices and use transactions such as credit card and phone call transactions on mobile devices, they can manipulate such data to steal money from the company in this manner. As mobiles have a constant built in billing system, malware can make use of this to steal large amounts of money from firms. An example of such a Trojan virus is called "Red Browser", that sends streams of text messages (each billed at around 5%) to all phones until the affected phone is disabled.

Mobile wallets are threatened by malware attacks, as making such important transactions over a mobile platform can be risky and susceptible to infection by malware.

Effects on Subscribers

In devices where bank account details and credit card information is stored and used by employees, identity theft can be a huge danger. The occurrence of malware using such information from personal computers to conduct bank fraud and other financial crimes has been common over the years. The sophistication of malware for mobile devices is not yet as high as that for personal computers, and so bank fraud and similar financial crimes are less likely to occur over mobile devices. However, this threat is not to be overlooked as technological advancements are taking place very rapidly in all sectors, including cybercrime.

Identity theft is still common among mobile malware attacks. They aim to steal information such as address book details to send unauthorized messages and mails to contacts under the identity of the user.

Effects on Mobile Operators

Upon infecting a mobile device, malware (or Trojan in particular) makes a connection to a command and control (C&C) site, which then uses the device as a botnet to spam other devices with SMS messages, or to target them on a DNS server or Dodos attack. This puts a great strain on the network's resources.

The strains on mobile networks may be such that the number of complaints by affected users may increase greatly, thereby jamming the network's help lines. Other than this, the time and effort the mobile companies are then required to put into the eradication of malware from devices puts a great strain on operator resources.

What operators can do to reduce these strains are use firewalls and policy engines to identify and block relevant C&C sites, which may harm mobile devices. However, increasing sophistication of malware can find a way around such hurdles and impact devices anyway.

Communication Based Attacks

Attacks can be based on means of communications and can affect mobile devices with just the press of a button or tapping an "Accept" icon in a device.

Attacks through SMS

If there is a network or mobile device flaw in SMS and MMS transmissions, malware attacks are made very simple. If a mobile phone has a development error and fails to manage binary text messages, the mere act of receiving a certain message or character can cause a DOS attack. The phone may restart, or it may to work properly. This problem is not only limited to SMS messages, but can also be caused through use of emails.

As is common in many countries, if a large number of SMS messages are sent through the Internet, this can cause an overload in the network and delay or cause failure in the deliverance of text messages. This lag in the network can be used to cause a DDOS (distributed denial of service) attack in large SMS infrastructures.

Other than simple SMS messages, MMS messages are also used for malware distribution. If an MMS is sent with a certain attachment that in reality is a virus, then upon opening the attachment, the virus will infect a mobile device. If the user does not open the attachment, the phone is safe. In the event that the device does become infected, it will automatically distribute the same MMS to all contacts in the phone's address book and cause harm to countless other devices.

Attacks through GSM

In a GSM network, malware attacks through breakage of the network's encryption. Once through that (it takes about 6 hours to complete this), the attacker can get through to all the unencrypted communication and information made by a Smartphone.

Attacks through Wi-Fi

Through close surveillance of a network's communication, an attacker can gain access to mobile devices that are connected to a Wi-Fi network. The Wi-Fi serves as the intermediary in connecting a device to the Internet. Hence, malware may be distributed through Wi-Fi networks. The WEP keys used for each Wi-Fi network are the same for all devices that are connected to the network and this is what creates the problem. However, use of the WPA security protocol can mitigate this risk of infecting mobile devices through Wi-Fi. However, even WPA has its limitations in the sense that small or simple keys can be subject to brute force attacks. Another version of WPA, i.e. WPA2 is said to be safe against such an attack.

Another way Wi-Fi networks are used to transmit malware is through creation of duplicate access points. If a dummy access point is created and the original network is not saved to the mobile device, the user can unknowingly connect to the dummy network, as it will have the same username and password. However, most smartphones have network memory that saves frequently used networks to the device and has the option to automatically connect to them, thereby reducing the effectiveness of dummy wireless networks gaining access to Smartphones.

The last way in which Wi-Fi networks can be used to infect mobile devices is through the installation of software installation scripts that are downloaded without the knowledge of the user. The device trusts the download to be from a trusted source and the installation of the SIS file ends up infecting the device.

Software & Application Based Attacks

Application based attacks are highly common these days among smartphones and tablet users. Whether you are using your mobile device for personal or official use, your mobile device is susceptible to malware attacks. This is not because you have done something wrong, but is in fact a flaw in the app market that sells you applications with malicious content.

The most affected platform is that of Google Android. For some reason, Google has faced high volumes of software attacks over the years. What is a software attack? It is similar to apps with malware in them. The reason why Google has been so greatly affected by this problem is due to the sheer amount of user base it enjoys. The more the Android platform is used and adapted to mobile phone developers, the more it is susceptible to malware attacks and phishing scams.

The significance of this problem is due to the rate at which hackers are attacking smartphones. This is said to be even greater than the level of hacking into personal computers in previous times.

The sophistication with which mobile phones are hacked into through applications is confounding. The applications gather information about the device and the personal information (as well as corporate data) that is stored in the mobile device. This information can then be used to conduct identity theft, financial fraud, sale of important and secret corporate data, company trade secrets to competitors and so on. Another threat of malicious applications is the spam they send to all the contacts in a device's address book, thereby infecting any other devices that receive and accept such messages and mails.

App stores at almost all operating systems are subject to clear and make available for sale and download apps with malicious content. This is because with each day, the sophistication and techniques used by cyber criminals who develop these apps are improving greatly. It is easier and easier for apps to get through security checks these days.

7 Effective Countermeasures

The Business world has faced an increasing need for mobility devices and software over the years, which have had both advantages and threats that come with integrated use of devices. The rising use of mobile devices has given rise to a new form of threats posed to the company and its people. Virus attacks and hacking are a common occurrence in enterprise mobility and needs to be addressed.

Mobile phones have evolved in sophistication and host multiple platform and executions nowadays. This makes them susceptible to several viral and hacking attacks that can harm the individuals who use them as well as the company they work for.

Companies should take threat management measures to mitigate the risks of using mobile devices around the workplace, especially for protection of important corporate data.

1. Encryption

For any corporate database, the protection of its contents is extremely critical. Whether the data is being used inside the company or through a mobile device away from the company's office location, it should be protected to the highest degree possible. The most basic and essential way to protect data is to code or encrypt it. Encrypted data can only be accessed using approved company interfaces, respective IDs and passwords for all employees. Any or unauthorized attempt to access key data can be identified immediately and offenders can be apprehended with ease.

2. Defining Duties

For every employee, there must be a concept of accountability and submissiveness to Furthermore, each employee should be assigned certain duties and access levels to corporate data. This makes it easier to detect misuse of databases in the organization and enhances the company's data security position. Monitoring the activities of workers against their defined duties can help the business determine when, if at all, a worker tries to access data he is not meant to have access to.

3. Protection Against Applications

Much like personal computers, mobile devices require security measures that protect them at least on the most basic levels against malware attacks. Application layer defenses and all kinds of activity monitoring etc. are to be used to minimize the threat of unwanted access to be gained into corporate mobile devices.

4. Sandboxing

For mobile devices, application sandboxing is a great Business solution. This is also known as containerization, and its Primary focus is to limit the environments in which certain codes can be executed. It is a process that isolates an application and prevents malware infections, or any forms of attacks from outside the app. It basically entails no outside access to the data and information inside the application that is sandboxed, which is much similar to children's sandboxes in which their toys are contained.

For companies: who want to implement this form of protection, sandboxing is possible in mobile devices. Other than this, the developers of apps themselves can wrap around their app certain security policies. The reason why sandboxing is not so popular, however, is that it is very complex to implement. The use of sandboxing can create even more security problems if the process is not carried out proficiently. If a sandbox has to contain in itself all the data that an application needs to execute, the applications that need to interact with each other can face problems doing so because of the sandbox.

5. Itinerary Recording

Having a faulty application in the system can pose threats similar to those from malicious content. Companies should invest in fault tolerant abilities and replication. For example, if malware is present in the system or the application, each computation is replicated numerous times to ensure accuracy of results. An agent is duplicated many times to ensure that not all of copies are affected by malware and the copies that go unaffected reach their destinations as required.

6. Application Encryption

If other methods of protection are failing, then encryption at the application level should be used as a backup. If a device is stolen or misplaced, data can still not be accessed without the key. *Masking* techniques can delete data from respective storage point's altogether to avoid security breaches. In large companies where data storage needs are extremely large, what *tokenization* can do is remove the most important data portions from the storage destination completely. If, in certain applications, such options are not provided, a business need not worry. A sound IT department or a hired developer can use an application and add the company's requirements to it. These add-ons are likely to work only on small-scale storage capabilities, however, not on large database requirements.

7. Mobile Device Management (MDM) Software

The ultimate solution is to invest in MDM. However, smaller businesses with limited budgets will have to go with the techniques mentioned above as in most cases, and MDM is not affordable or feasible for a company to implement.

CHAPTER 12

Mobile Business Intelligence

What is Mobile Business Intelligence?

Let us start with a quick review of the definition of Business Intelligence or BI. Gartner gives the following definition of BI in its online IT Glossary: "BI is an umbrella term that includes the applications, infrastructure and tools, and best practices that enable access to and analysis of information to improve and optimize decisions and performance."

Gartner IT Glossary

Adding the term mobile simply implies the access to similar tools on a mobile platform. Mobile Business Intelligence serves to provide users and devices access to visually engaging information through analytic tools.

The mentioning of users and devices separately was intentional as there is a difference between the two. Yellow Fin, a Business Intelligence vendor, explains that there is a category of users who use multiple devices; these are referred to as device mobile. They can access the Mobile BI through smartphones, tablets, laptops or desktop PCs. Then there is the second typical category of mobile users who use a mobile on the go. These users are physically mobile and utilize the Mobile BI services from multiple locations. However, the categorization is for the ease of understanding. In reality, it is common to see these two categories overlap.

The Mobile BI makes it possible for the user to access complex analytic tools (I.e. that demand high computing capacity) on their mobile devices. Mobile BI emphasizes on providing visual reports in the form of graphs, bar charts and trend lines on the interface/ dashboard.

According to research firm Aberdeen Group, the benefits users seek from Mobile BI include greater flexibility and greater ability to respond to customers. Market research reports confirm rise in usage of mobile platform to access BI analytics. As this prediction materializes companies would design BI apps that provide an optimal experience in mobile devices alongside PC desktops. The trend is known as 'device independence'.

A Brief History of Mobile BI

The advancement in desktop BI solutions has been faster than that of mobile BI solutions for the simple reason that earlier mobile devices could not support complex applications. An academic research points out four reasons for the slow development of mobile BI.

- Poor resolution: The resolution in earlier mobile phones was insufficient to provide attractive BI visuals
- Small screen size: With an average diagonal screen size of 12 inches, a user could not get a reasonable view of BI charts on mobile devices
- Small memory: The search found that the average range of RAM was between 256-52 MB, which is very minute compared to desktop counterparts
- Low processing power: The mobile platforms did not have the processing power to smoothly run heavy

software. BI tools are usually very complex and may slow down systems with lower processing capacity.

However now that the technological landscape has evolved, the mobile devices have come close to desktop PCs in multiple aspects of functionality.

While the history of BI goes back to 1958, the concept of Mobile BI is relatively new. Apple's iPad launch in 2010 was the starting point for Mobile BI. The tablet's PC-like-functions and wide display touchscreen made it ideal for BI apps. Since its launch, the tablet PCs have received massive acceptance. According to Business Insider, the supply of tablets will beat supply of PCs within two to three years.

Additionally, the first tablet BI application was given by Micro Strategy in the same year, i.e. 2010. The app was compatible with iPad and iPhone. The app was available through the Apple store and was priced at 550 to 2,000 USD. The app allowed users to custom design dashboards and use the iconic touch and swipe feature of Apple devices.

Looking at the positive response other BI vendors also entered mobile BI market. Since then, the Mobile BI market is innovating at a faster pace to keep up with the requirements of the users on the go. The Business Insider report suggests that in coming years, most Internet access will be gained through mobile devices and not static/ fixed devices. While this may mean survival crisis for desktop PC manufacturers, it signals rising competition for mobile device vendors and mobile device app vendors. One should expect the Mobile BI market to experience significant changes in the coming years.

Moving from passive to active BI & from push to pull reporting

Initially analytic tools comprised of scheduled reports. These reports were submitted on a regular basis (daily, weekly or monthly) and were highly role-specific, for example, a daily absenteeism report sent to the HR manager or a weekly sales expense report to a finance manager. The system was simple and did not entertain any queries. This approach was regarded as passive or push reporting.

As technology changed, active mobile BI reporting was introduced. This reporting approach allowed the user to submit questions to the app and receive customized reports. The BI apps became more responsive. It is also referred to as pull reporting since the user can pull required information from the central data warehouse. For example, the HR manager may ask for department wise segmentation of absenteeism report.

Most companies use a hybrid approach towards mobile BI reporting. While executives have access to active and passive mobile BI apps, the factory staff usually receives passive apps only. The decision of who receives what information is dependent on the person's managerial level and related information requirement.

Purpose Built Mobile BI Apps

A look at the companies that have deployed Mobile BI can help gain insight about the usefulness of this technology tool. Given below are the details of Mobile BI deployment at Sonic Automotive Inc.

Mobile BI making its way at Sonic Automotive Inc.

Sonic Automobile Inc. is one of the leading players in the US Automobile industry. Bruton Smith, who is also the Chief Executive Officer of Sonic Automobile Inc., founded the company in 1997.

Company overview as of 2013	
Company Ticker	NYSE: SAH
Industry	Automobile
CEO	Bruton Smith
Dealerships	111 approx.
Achievements	Fortune 500
	Russell 2000 index

The company has grown from a 7-dealership network in 1997 to a countrywide network of 111 dealerships as of the year ended December 2012. As the network grew larger, the need for efficient communication and fast decision-making increased.

The company's CEO had recently purchased an iPhone and was quite impressed by its innovative technology. The management started looking into opportunities to achieve operational excellence. They were looking into areas of business intelligence but the design of the iPhone did not support the desired functionality. Sonic entered into talks with technology companies including Apple in 2010. The same year iPad and the first Mobile BI app were launched. Sonic saw value in this and opted for mobile BI.

Sonic wanted to bring the time taken to buy a car down from four hours to one hour, says Byrd. The company wanted sales staff to access relevant data so that better decisions are made based on consistent and updated information that is free from any error caused by subjective assessment. The company wanted to deploy something that would create 'value' said the CFO.

The mobile BI solution: In 2010, Sonic deployed Micro strategy's Mobile BI solution on iPads to fulfill its requirements. Almost 250 people including regional managers and top executives were given iPads. The cost incurred per unit was 2,500 USD (Excluding installation cost). The deployment plan included setting up six access points at every dealership outlet. The mobile BI tool showed results in the form of visually compelling reports specially designed for the sales staff. Later on, the Mobile BI was combined with another app named FIRE (F&I Reporting Engine) to produce BI reports with enhanced visuals.

Challenges: The deployment was not an easy process. It required a deep cost and benefit analysis of the situation. Despite the fact that there was no guarantee of success, it was decided the chance was worth the taking. Byrd regarded the Mobile Data Management (MDM) and provision of web access points as the main challenges in deploying the technology. The management had to make sure that everyone was at the same level i.e. using the same version to ensure streamlined information access. This called for regular updates. Bandwidth upgrades were made regularly so that all employees could access the Internet without hindrance and delay.

The results: The results were promising. Firstly, company revenues improved. As evident from the graph above, Sonic saw a rise in its year-to-year revenue in in the first year after deploying mobile BI. The increase was approximately Twelve percent. [2011 revenue -2010 revenue) /2011 revenue]. Secondly, the intended value was achieved in the form of fast and informed decision-making. As a result, the company received the 'Best Practices in Business Intelligence' award in 2012 for its fast adoption of mobile BI.

However, it does not stop here. As of 2013, the company has given 7,000 iPads to its employees to benefit from mobile BI. The focus has shifted from sales to all functional departments.

The case of Sonic Automobile Inc. deploying Mobile BI as soon as it was introduced in the market shows that the company had a futuristic approach towards achieving business excellence. This, along with many other success stories, reflects the benefit of properly implemented technology.

Web Applications vs. Device Specific Applications

The applications provided by mobile BI vendors can be either web applications or device specific applications. Let's have a look at the two forms in detail.

Web Applications

General definition - Web based apps are accessed through a browser and there is no need to install the app on the device for use.

Mobile BI context - Web based BI apps are those analytic tools that are accessed online. The benefits of having a web based mobile app are as follows:

- These apps are compatible with all browsers offering convenience to the user
- They are easier to develop and there is no need to cater to device specifications since the app is not installed on the mobile device
- They are less costly to develop. No device specifications reduces development cost
- The HTML5 apps show features close, if not 100% similar, to native apps
- All users are on the same version as upgrading is done at the browser level

However, web based apps have some drawbacks as well:

- The offline performance of web based mobile apps is not as effective
- The functionality is not Up to par with native apps
- The web based apps do not provide optimal user experience as their design does not incorporate differences amongst mobile devices
- The app does not have access to all the hardware and software features of the device
- There is no centralized source for accessing the app

Native/ Device Applications

General definition- Native apps or device apps need to be installed on to the device for use. They differ according to the mobile devices i.e. mobile or a tablet.

Mobile BI context- Native apps reside on the device and provide the user with the following benefits:

- These are standalone apps, which can work even without a browser
- Once installed these apps provide better functionality and features than web based apps
- In absence of internet access the functioning is not compromised
- The apps integrate with all the hardware and software features of the mobile device to produce better reports
- The app provides an optimal user experience as the design incorporates mobile device specifications
- These apps are available at secure and centralized sources i.e. Apple's App store offer all native apps for iPhone and iPads.

The drawbacks of adopting this type of App for mobile BI usage are:

- Due to the high level of specifications, native apps are costly to develop. Each mobile device has its own specifications and hence a separate app design
- The upgrading of the app is at the discretion of the user. Because of this, at a given time different users will have multiple versions of the same app
- Lack compatibility with all mobile devices

Choosing between Web-based and Device-based mobile BI apps

Having stated the benefits and drawbacks of the two types of mobile BI apps one may ask which one to choose for their company. The answer depends on the company's information requirement and resource availability. Companies with substantial IT and financing resources and a focus on efficiency go with the device based/ native apps. Native apps produce good results if most of the company users have same mobile platform. However, the decision is often made in favor of a hybrid app i.e. which is available in both forms. An example of a hybrid mobile BI vendor is Jasper soft. Jasper soft is a mobile Business Intelligence solutions firm that offers both BI solutions to meet the rising demand for hybrid apps.

In light of the 2013 market research by Aberdeen Group, companies prefer native apps over apps when it comes to mobile BI. The results showed that Fifty-eight percent of the best performing companies had a native mobile BI app. The improvement by the use of native apps was visible in the form of rising operating profit. Plus the native app users enjoyed faster access to timely information as compared to web based app users.

The Demand for Mobile BI

Research suggests that mobile BI users are now much more concerned about ease of use relative to functionality. The rise in smartphone sales and internet penetration all contribute to rise in demand for mobility at all levels, and BI is no exception.

Mobile BI User Segments

According to an article published in the World Academy of Science and Technology journal, the demand for Mobile BI is driven by four user segments i.e. Executives, Field Workers,

Business Analysts and Clerical staff. Each user segment has access to either passive Mobile BI or active mobile BI.

- Executives: The top management requires full-time access to information about Key Performance Indicators (KPIs). The active reporting approach is employed here. The BI is responsive and can generate according to user queries.
- Field Workers: Theses includes field workers like sales staff. They need information specific to their field. A hybrid of active and passive reporting approach is employed in this case. The information the sales staff can access is less critical compared to the information available to executives.
- Business Analysts: These can be employees from any department. They derive information from mostly passive BI applications.
- Clerical staff: Though the use of Mobile BI at this level is very limited but it can enhance productivity of clerical staff. They also rely on static reports generated by passive BI applications.

Usually the first two categories are kept in focus while determining the demand for reporting approaches and the type of Mobile BI apps to install. An online study titled 'Wisdom of the crowds: mobile computing/ mobile business intelligence survey study' shows that executives have the highest level of cultural preparedness to accept mobile BI. This is because this user group has the greatest exposure to desktop BI apps in a company, so changing the platform is no big deal. It is followed by sales and marketing staff, which claimed that it is 'somewhat' prepared to accept mobile BI.

Industry demand

The need for Mobile BI is not consistent through industries. Some industries require Mobile BI while others do well with desktop BI. According to an online research survey the demand for Mobile BI is the highest in the consumer goods industry (the sample included four other industries i.e. Retail & Wholesale, Health care, Financial Services and Government firms). Eighty percent of the respondents believed that Mobile BI was critical for the consumer goods industry.

Market statistics

The figures from a 2012 report by The Data Warehousing Institute (TDWI) regarding the mobile BI trend brought forward the following important findings:

- In the next 12 months Sixty-one percent respondents plan to increase usage of mobile BI seventy percent
- of retailers rate mobile BI as an important part of the firm's business intelligence
- Past data suggest a rise in Mobile BI users although the current number of users is not so high
- Retailers with existing BI system will be successful when the BI tools extend to mobile devices

These stats confirm a rise in demand although the trend has not penetrated deep into the market yet. The lack of devices with widescreen displays and required computing capabilities could be a possible cause for the low penetration in the past, as mentioned earlier. This has left a lot of room for growth. Experts at Forrester Research believe that the growth in widescreen Smartphones and tablets, and availability of Mobile BI solutions will collectively boost Mobile BI usage.

In order to be considered successful, a feature of Mobile BI tools must be 1) optimized and 2) responsive. An analytical tool having the capability to present analysis reports on mobile devices is not enough. The very design of the tool should be made keeping in mind the end user mobile devices. Only then can be the tool yield optimal results. Users are not looking forward to access only; rather they want optimal analytic experience. In this market, the demand is largely captured by Apple. Apple has been leading the tablet markets and the tablets have been leading Enterprise mobility.

Future prospects of Mobile BI

The future of Mobile BI is bright if we look at the adoption rate of this technology and the demand growth graphs. According to research firm Gartner, the Mobile BI market is expected to witness growth alongside innovation in years to come. The industry that was stagnant for numerous years will now grow by double digits. Mobile BI user base will increase by Fifty percent by 2015 says Gartner. This increase is an outcome of the enterprise mobility trend. More smartphones, greater and faster web access and need for fast decision-making have given a push to this market.

The demand for better and innovative BI tools will surge as well. Gartner predicts that in the near future the market will receive Mobile BI dashboards with panoramic analysis i.e. data will be pulled from multiple sources and integrated in the report to give powerful insights. Data from industry, economy, and competition, business partners and from within the organization will be compiled to give a bigger picture of events. The future is all about integration and synchronization. The Mobile BI apps of the future will have advanced features like spoken queries/reports and video reports.

Business Benefits of Mobile BI

The benefits of Mobile BI are fully visible when it is adopted on a large scale in the company. However, all things start small. Therefore, the IT department must ensure fast adoption rate to avail the following benefits:

Faster decision-making: Research by Aberdeen Group found that decision making in organizations with mobile BI platform is six times faster than an organization with no mobile BI. Efficient decisions can save up on the opportunity costs incurred due to delayed response. The Mobile BI through its visual display serves as a platform to support not only faster but also effective decisions.

Sustaining competitive advantage: According to an academic research, Mobile BI can serve to provide a competitive advantage. This is how it works; consider two competitors who do not compete based on resources (i.e. Human capital or financial capital) rather the competition is driven by efficient and effective decision-making. Timely decision-making is a tool for competitive advantage as it helps in building a proactive stance towards market changes.

Time saving: Previously, employees had to go through piles of data to find the information they were looking for. This was a time-consuming and stressful task. The algorithms working within the Mobile BI tool make it possible to extract relevant information from data sources and present it in a visually appealing and useful manner without exhausting the end user.

Higher employee engagement: Mobile BI experience offers a higher degree of employee engagement as the display format is more visually appealing. Gone are the days when executives and sales staff had to go through boring standardized multi-page reports for making a decision. Because of this aspect Mobile BI, result in higher employee productivity. Better outputs are achieved in the same or less amount of time.

The analysts believe that giving front-line staff access to relevant data inculcates a feeling of autonomy amongst them increasing their confidence level. This in turn leads to better decisions.

Customer Satisfaction through smooth information flow: Since Mobile BI equips the front-line staff with relevant information; the staff can serve the customers better. Firstly, it reduces service delays since the front line staff has real-time knowledge about inventory levels. Secondly, even if there happens to be a delay because of some uncontrollable external factor the staff can convey the information to the customer. Consumer behavior research confirms that customers tend to be comfortable with informed wait. The smooth information flow from company data center to front-line staff and customers, leads to high levels of customer satisfaction.

While it may seem as if achieving all the above benefits are near to impossible, this is not always the case. The key is to deploy for the purpose of creating and enhancing value not because the technology is the 'in' thing. Having a well-defined and goal-oriented approach towards deploying this technology can yield maximum advantage. With the right mentality, technology deployment can do wonders, literally.

Security Concerns Regarding Mobile BI

Wherever there is access to company data there are security concerns. Critics have long regarded technology as increasing threats to corporate data. As technology advocates enhanced democratization and mobility of data, the task of keeping the security of the data intact becomes harder.

However as stated in earlier chapters, security challenges are inherent to all user devices let it be PCs or smartphones or tablets. Avoiding the mobile BI usage altogether is not the solution as experts regard mobile BI as a source of competitive advantage. The need is to have a firm action plan in place to deter security threats linked to compromised devices.

The common measures to deal with security challenges include:

- Managing access - Access to information should controlled by setting session time-out restrictions and applying passwords.
- Managing storage - In case of web based Mobile BI apps there is a lesser need to worry about data security since no information is stored on the device. In the case of native mobile BI apps, the device bears all the data. Encryption can serve to protect stored data.
- Remote data wipe - In the event of a stolen or misplaced device there should be a central source i.e. IT department, from where all data can be wiped.
- Device Security - The Company should encourage users to install anti-theft software or provide them. These allow the user to access the phone even when stolen. For example, Norton Anti-Theft software covers multiple mobile devices. The software can track the device, block access to information and even take pictures of the thief to assist in retrieving the device.

The key to successful security policy is to keep it flexible yet firm. Rigid policies often affect user experience due to excessive restrictions. The market is filled with security solutions that offer varying features, the challenge it to find the solution that fits with the company requirements.

CHAPTER 13

The Future of Enterprise Mobility

Look Out for Key Technology Changes

"Once a new technology rolls over you, if you're not part of the steamroller, you're part of the road."

-Stewart Brand, American Writer

The IT industry has experienced a rapid and intense change over the years. A few decades ago, PC used to dominate the market, and then came laptops and cell phones to steal the thunder from PCs. Now we have tablets and smartphones. The industry moved from predominantly static products into mostly mobile products.

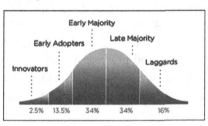

Figure 6: Technology Adoption Lifecycle

Typically, whenever there is a technological change, the industry players become segmented based on adoption speed. The model referred to as Technology Adoption Lifecycle, shown in figure 1, is a widely accepted concept.

However, if we look into the nature of the IT industry there has always been a never-ending race of innovation and breakthrough technologies. The speed with which old technologies are rendered obsolete is amazing. This implies the high cost of being a late adopter, let alone being a laggard. Laggards have no chance of surviving in the long term.

While the innovators set the benchmark, the early adopters follow suite. Blackberry and Apple emerged as innovators in different times, with regard to mobile technologies. Companies like Sonic Automobile and Intel were amongst the early adopters of different types of mobility trends.

Companies with active and future-oriented leaders (CIOs specifically) have the ability to differentiate a trend from a fad. This ability allows them to think fast and act faster, hence overcoming costs of delayed response. The cost of delayed action towards adopting new technologies can be in the form of lost market share, lower operational efficiencies and lesser employee productivity.

This does not mean that one invest millions in every new technology just because it has a 'latest' tag attached to it. The aim is to look for technology changes that will create value for the company and produce substantial outcomes. For this purpose, it is necessary to have knowledgeable leaders on board, who can lead the company in the right direction.

An Outlook for Employees

As mobility platforms tend to evolve over time, it is necessary for a company to keep its employees updated and trained to make full use of the opportunity technology advancement offers. The following aspects require consideration.

Keeping all employees at the same level

On the company's part, it is essential to keep everyone on the same level as the company makes technology shifts. All departments should ideally have the same technology, though the extent of usage may differ. The goal is to reduce the chances of technology disparity across departments that could affect employees on rotational jobs.

For instance, one department may predominantly be using an old version of a software or OS while another may have updated to the latest version. The difference in technology adoption rates can make office tasks hard for the employees when shifting departments. This problem is aggravated when there is a high BYOD trend and when most of the corporate apps are device specific. This delegates the responsibility of upgrading to the employee who may or may not upgrade instantly when a new version arrives.

This problem can be overcome by another growing trend toward Corporate Owned, Personally Enabled (COPE). According to an article[22] published in The Guardian, the core purpose of both BYOD and COPE remains same: enabling mobility. The only difference is that COPE comes along with implementation and control benefits for the company. While the employees can use the device for work and corporate purposes (the same as in BYOD), the company can easily manage and upgrade the devices so users have access to consistent information.

Training for better customer service

Today's customers are smarter as compared to those in the past. The mobility trend has not remained confined to the companies;

---

[22]   Retrieved from http://www.theguardian.com/media-network/media-network-blog/2013/apr/24/corporate-owned-personally-enabled-cope-byod

it has mobilized customers as well. With a single touch, customers can access all relevant information about products, promotions and sales. A smart customer can only be served well if the employee (i.e. front line staff) is smarter and more knowledgeable.

In this regard, the marketing department faces challenges to provide the demanded level of customer service. The mobile customers prefer more targeted marketing campaigns. Consider the following:

- Market reports suggest that Smartphone users spend a significant amount of time on online socializing activities.
- Com Score suggests a rise in the level of online retail shopping by mobile customers. The growth rate ranges from Forty-five to eighty percent for different types of online retail categories
- A Google sponsored research[23] shows that as much as Seventy-nine percent of smartphone owners are online shoppers. An increasing number of Smartphone users rely on their devices while making shopping decisions.
- Talking specifically about Asia Pacific customers, stats[24] suggest a rapid shift from traditional marketing communications channels (Email, telephone and salesman) to new platforms (Mobile devices, social media and instant messaging). The trend will grow faster in coming years.

---

[23] M.A.R.C. Research & Google Shopper Council. (2013). "Mobile In-Store Research: How in-store shoppers are using mobile devices". Google. Retrieved from http://googlemobileads.blogspot.com/2013/05/understanding-smartphone-use-in-stores.html

[24] Economist Intelligence Group. (2012). Service 2020: Megatrends for the decade ahead. BDO Australia. Retrieved from http://www.bdo.com.au/data/assets/pdf_file/0008/142892/Service-2020_AUS-Megatrend-Eight_FINAL.pdf

- Customers expect round the clock 24/7 availability of customer services staff
- Mobile device users want same website loading speed as the desktop counterparts. Customers may never visit or recommend sites that load slowly on mobile devices.

All of these factors suggest that mobile customers are growing in number and in shopping activity as well. Keeping a record of mobile customers, their usage patterns, socializing activities and shopping preferences is more important than ever if companies intend to retain customers.

Keep an Eye on Areas of Interest in Your Business

Various forms and areas of mobility have been mentioned in the previous chapters. The breakdown can be made according to:

- Functional departments: Company may allow mobility measures in one department, e.g. sales and not in another based on need.
- Managerial level: Top management and front line staff may be allowed to use mobile devices while others may be required to use company PCs.

Whatever the reason, mobility can be enterprise wide or focused on a few individuals or departments.

Looking ahead, mobility measures should be extended based on industry type. As stated earlier some industries, e.g. consumer goods industry, have a huge need for mobility. Industry dynamics serve as a guideline for determining which direction to take with mobility. Even within a specific industry, certain areas demand more mobility than others do. For example, the shop floor attendants in retail industry need to be well equipped with all product details (own and competitor's both), as a large portion of customers that step into the store have product details and price comparisons in their handheld device.

Coming back to industry practices, there are many firms that set these industry trends in the first place. They initiate the debate regarding whether to mobilize or not. These pioneers are often those companies that have vast experience and reliable business insight with little if any financial constraints. They can serve as an example for other players in the market.

For small companies the game is not so simple. If the cost of failure is high, the reward of success is significant as well. They can start by performing a cost-benefit analysis of potential areas for implementing mobility measures.

Update and Integrate Your Operational Model Frequently

As technology evolves, it is important that business models evolve as well. Who knew that a socializing website developed by some Harvard student would become a million-dollar commercial platform? The thing is that industries never remain stagnant, business processes evolve and so do business models. The boom in social media and the shift from PCs to mobile devices necessitated the updating of business models. Customer service and marketing channels all require constant updating to ensure the survival of a business in today's worlds.

The mobility of both external customers and internal customers (i.e. Employees) has repercussions for the corporate business model. Impact of mobile employees asks for reviewing:

- Corporate information channels (from fixed infrastructure to mobile infrastructure)
- Communication medium (from emails to instant messages between employees)

- Support systems (from in-house servers/ IT department to scalable/ third party cloud vendors)
- Control mechanisms (rigid to flexible- BYOD or COPE)

The rise in mobile customer demands investment in the following aspects:

- Developing apps for assisting with shopping
- Optimizing websites for mobile users alongside fixed device users
- Offering 24/7 customer care center to attend to queries

Stated above are a few points that can be reviewed as the company ventures on the road the two mobility.

Ride the Mobility Wave with Poise

Here are some tips that can help in dealing with the growing emphasis on enterprise mobility:

- Do Your Research: Before going mobile, perform thorough market research of mobility consultants and mobility solution vendors. Select the solution that best fits with your company based on company size, industry type, number of mobile employee and mobile security preferences.
- Start Small: As you begin to launch mobility movement in your company it is important to start small. This will be beneficial in 1) identifying possible bottlenecks in the system, i.e. slow speed or poor access and 2) gathering user feedback on user experience, i.e. interface and policies.
- Define policy: Mobility is not synonymous to letting people do work as they find productive. Usage policies specifying choice of device, access control, data management rights all should be conveyed to employees willing to register for mobility program.

- Stay focused: It is important to stick to the purpose of going mobile in the first place. Usually it is to enhance productivity. Just like a mission statement, a mobility statement should be made and circulated to every employee so they know what to expect and what is expected. A well-stated mobility purpose statement can help by acting as a benchmark to evaluate performance.
- Calculate ROI: No one wants to start the mobility movement if the ROI was poor. The reality is companies invest where they see value. The value may or may not be tangible, making it hard at a later stage to figure out how worthwhile the investment is. ROI measures need to be in place before implementing BYOD or COPE for ease of evaluation purposes.

These simple steps provide a genuine guideline while embarking on the journey to mobilizing your business. Attention to detail holds great importance in this shift, as it does in any other transformation.